Contents

MyMaths

for Key Stage 3

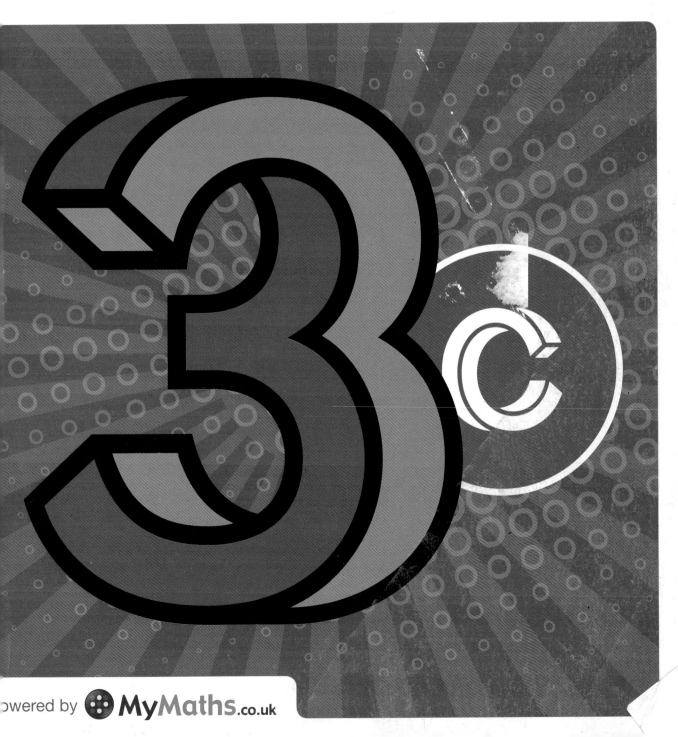

powered by **MyMaths**.co.uk

Acknowledgements

Although we have made every effort to trace and contact copyright holders before publication this has not been possible in all cases. If notified, the publisher will rectify any errors or omissions at the earliest opportunity.

p5: chungking/Shutterstock; **p8**: Carsten Reisinger/Shutterstock; **p11**: Royal Belgian Institute of Natural Sciences; **p24**: Bryon Palmer/Shutterstock; **p34**: itsmejust / Shutterstock; **p39**: Yuriy Boyko / Shutterstock.com; **p40**: Alan Bailey/Shutterstock; **p42**: Portrait of Albert Einstein, c.1947 (b/w photo), Turner, Oren Jack (fl.1947) / Library of Congress, Washington D.C., USA / The Bridgeman Art Library; **p43**: JOSE ANTONIO PEÑAS/SCIENCE PHOTO LIBRARY; **p61**: Cowardlion / Dreamstime; **p74**: Andrey Popov/Shutterstock; **p76**: Frontpage / Shutterstock; **p79**: Roberto Fumagalli / Alamy; **p82**: Serg Dibrova/Shutterstock; **p90**: Becky Stares/Shutterstock; **p95**: Stephen Clarke/Shutterstock; **p98**: Kostenyukova Nataliya / Shutterstock; **p135**: Amy Johansson/Shutterstock; **p137**: George S de Blonsky / Alamy; **p138**: (t) Kevin Britland/Alamy; **p140**: Tony Bowler/Shutterstock; **p151**: INTERFOTO / Alamy; **p161**: Brian A Jackson / Shutterstock; **p162**: kamnuan/ Shutterstock; **p164**: Gina Smith/ Shutterstock; **p166**: Christopher Rennie/Robert Harding World Imagery/Corbis; **p169**: Azat1976 / Shutterstock; **p181**: Eldad Carin/ Shutterstock; **p187**: matteusus/iStockphoto; **p191**: ClickHere/Shutterstock; **p199**: Vadim Sadovski/Shutterstock; **p203**: Georgios Kollidas / Shutterstock; **p215**: DEA / A. DAGLI ORTI/Getty Images; **p218**: dpa picture alliance archive / Alamy; **p236**: U.S. DEPT. OF ENERGY/SCIENCE PHOTO LIBRARY; **p237**: Evgeny Dubinchuk/Shutterstock; **p243**: Alex Emanuel Koch/Shutterstock; **p265**: bluehand/ Shuterstock; **p273**: Vtls/Shutterstock; **p288**: Gjermund/ Shutterstock; **p291**: Justin Kase zninez / Alamy

Case Studies:

Online Creative Media/iStockphoto; OUP.Corel; Dean Turner/ iStockphoto; Scol/Dreamstime; Stefan Klein/iStockphoto; Mary Evans/iStockphoto; Mikhail Kokhanchikov/ iStockphoto; Kmitu/iStockphoto; Juri Samsonov/Dreamstime; Terraxplorer/iStockphoto; Timothy Large/iStockphoto; Martin Applegate/iStockphoto; Lisa F Young/Dreamstime; Blackred/ iStockphoto; Robert Mizerek/Dreamstime; Jf123/Dreamstime; Andy Brown/Dreamstime; Heather Down/iStockphoto; Hulton Archive/iStockphoto; James Davidson/Dreamstime; Brianna May/iStockphoto; Stefanie Leuker/Dreamstime; Viktor Pravdica/Dreamstime; Karam Miri/iStockphoto; Giorgio Fochesato/iStockphoto; Greenwales/Alamy; Mehmet Salih Guler/iStockphoto; Mediagfx/Dreamstime; Dusan Zidar/Dreamstime; Achim Prill/iStockphoto; Lucian Coman/ iStockphoto; Sierpniowka/Dreamstime; Felinda/iStockphoto; Robyn Mackenzie/Dreamstime; Elena Elisseeva/Dreamstime; TT/iStockphoto; Fernando Soares/Dreamstime; Jan Rihak/ iStockphoto; Gerald Hng Wai Chuin/Dreamstime; Hanquan Chen/iStockphoto; Yakobchuk/Dreamstime; Valentin Garcia/ Dreamstime; fotolinchen/iStockphoto; Dmitry Kovyazin/ Dreamstime; Rixie/Dreamstime; Feng Yu/Dreamstime; Valentin Garcia/Dreamstime; John Archer/iStockphoto; Christophe Test/Dreamstime.

Artwork by; Phil Hackett, Erwin Haya, Paul Hostetler, Dusan Pavlic, Giulia Rivolta, Katri Valkamo & QBS.

MyMaths for Key Stage 3 is an exciting new series designed for schools following the new National Curriculum for mathematics. This book has been written to help you to grow your mathematical knowledge and skills during Key Stage 3.

Each topic starts with an Introduction that shows why it is relevant to real life and includes a short *Check in* exercise to see if you are ready to start the topic.

Inside each chapter, you will find lots of worked examples and questions for you to try along with interesting facts. There's basic practice to build your confidence, as well as problem solving. You might also notice the **4-digit codes** at the bottom of the page, which you can type into the search bar on the *MyMaths* site to take you straight to the relevant *MyMaths* lesson for more help in understanding and extra practice.

At the end of each chapter you will find *MySummary*, which tests what you've learned and suggests what you could try next to improve your skills even further. The *What next?* box details further resources available in the supporting online products.

Maths is a vitally important subject, not just for you while you are at school but also for when you grow up. We hope that this book will lead to a greater enjoyment of the subject and that it will help you to realise how useful maths is to your everyday life.

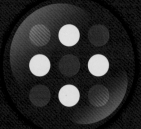

1 Whole numbers and decimals

Introduction

On the 3rd January 2004 the Mars Exploration Rover landed on the Martian surface within 200 m of its target. Navigators used radio signals sent by three antennae spread around the Earth's surface to control the flight of the spacecraft. An error in their distance calculations of even 5 cm on Earth would have led to an error on Mars of over 400 m.

In a journey of 300 million miles the navigators and scientists needed to work to incredibly high levels of accuracy.

What's the point?

Scientists, engineers and nurses need to know how accurate their measurements must be, otherwise there could be disastrous consequences. An understanding of rounding is vital in all walks of life.

Objectives

By the end of this chapter, you will have learned how to ...

- Round numbers to a given number of significant figures.
- Use rounding to make estimates.
- Find the upper and lower bounds of a calculation or measurement.
- Use prime factors to find the HCF and LCM of pairs of numbers.

Check in

1. Round each of these numbers to the nearest
   i 10 **ii** 1 dp **iii** 3 dp
   a 12.9725 **b** 342.91378

2. Write the value of
   a 3^3 **b** 2^5 **c** 10^6 **d** 5^4

3. Write all the factors of 72.

4. List the prime numbers less than 30.

Starter problem

If you fire a projectile such as rocket or a missile at an angle, you can calculate how far it will travel. You call this distance its range.

Range $= \dfrac{v^2 \sin (2\alpha)}{g}$

α = angle at which you fire your rocket

v = starting speed (velocity) of your projectile

$g = 9.8\,\text{m/s}^2$

At what speed and angle should I fire my projectile so that it travels an exact distance of 100 km?

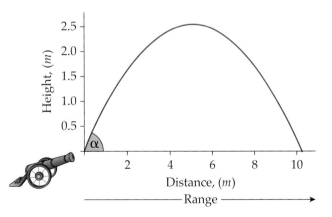

The position of a digit in a number is called its **place value**.

> ● The first non-zero digit in a number is called the first
> **significant figure** because it has the greatest value.

Significant figure is often abbreviated to sf.

1st significant figure is 2 (20)   3rd significant figure is 4 (0.4)

23.456

2nd significant figure is 3

You can use significant figures to **round** numbers and **estimate**.

Example

Round the numbers **a**   1234.5678 and   **b**   0.0012345678 to
i   1 sf   **ii**   3 sf   **iii**   5 sf.

- - -

a   **i**   The first significant figure is 1.   **b**   **i**   0.0012345678 = 0.0012 (1 sf)
1234.5678 = 1000 (1 sf)

  ii   The third significant figure is 3.   **ii**   0.0012345678 = 0.00123 (3 sf)
1234.5678 = 1230 (3 sf)

  iii   The fifth significant figure is 5.   **iii**   0.0012345678 = 0.0012346 (5 sf)
1234.5678 = 1234.6 (5 sf)

When you round a number you always look at the next digit. If it is 5 or more then you round up, if it is less than 5 then you round down.

Example

Estimate $\dfrac{23 \times 189}{3.9}$

- - -

Round each number in the calculation to 1 significant figure.

$$\frac{23 \times 189}{3.9} \approx \frac{20 \times 200}{4} \approx \frac{4000}{4} \approx 1000$$

Example

Siobhan buys and sells cheese. In October she buys 5252 cheeses at a price of £3.29 a cheese. Estimate the amount of money Siobhan spends on cheese each day.

- - -

Write the problem as a calculation. $\dfrac{5252 \times 3.29}{31}$

Round each number to 1 sf. $\dfrac{5000 \times 3}{30} \approx 500$

Siobhan spends roughly £500 per day on cheese.

Exercise 1a

1 Round each of these numbers to
   i   1 sf   ii   2 sf   iii   3 sf
   a   4219   b   6788
   c   1954   d   4003
   e   24 286   f   37 405
   g   4698.7   h   2854.495
   i   569 867   j   32 678 235

2 Round each of these numbers to
   i   1 sf   ii   2 sf   iii   3 sf
   a   5.467   b   7.394
   c   8.307   d   8.194
   e   4.5146   f   28.5251
   g   217.3474   h   0.8917
   i   0.034 57   j   0.007 394
   k   2.000 367   l   0.000 489 1

3 Use a calculator to work out each of these
   calculations and then round your answer
   to 3 sf.
   a 123.54×11.89
   b $2 \div 3$
   c $2.457 \div 145.89$
   d 0.0175×82.37

4 Round each of the numbers to 1 sf and
   then estimate the answer to each of these
   calculations.
   a $\dfrac{54 \times 3.7}{17.3}$ b $\dfrac{16.78 \times 1.9}{8.27}$
   c $\dfrac{2334 \times 9.8}{376.2}$ d $\dfrac{2.8 \times 47}{4.967}$
   e $\dfrac{28 \times 2.4}{0.57}$ f $\dfrac{20.99 \times 0.04}{1.78}$

Problem solving

5 For each of these calculations work out
   i   the exact answer to 2 sf
   ii the answer if you round the numbers to 2 sf after each step of
       the calculation.
   a   Boris buys 14 packets of sweets at £1.89 per packet.
       Altogether he has 209 sweets. How much does each sweet
       cost in pence?
   b   Sarah drives 1417 km each month. She notices that her van
       travels 9.2 km for each litre of petrol. A litre of petrol costs
       142.9 p. How much does Sarah spend in pounds on petrol
       each year?
   c   Axel the cat eats 325 g of tinned food each day. He also
       eats 230 g of biscuits each week. A 400 g tin of cat food
       costs 55 p. A 1.5 kg bag of biscuits costs £7.89. How much
       does Axel cost to feed in a year?

Did you know?

In 2011 there were
28 500 000 cars on the
roads of the UK.

6 Hector and Giles are rounding the answer to a calculation.
   Hector rounds the answer to 2 sf and gets an answer of 1.4
   Giles rounds the answer to 3 sf and gets an answer of 1.40
   Hector says that both answers are the same.
   Giles thinks that the answers are different.
   Explain and justify.

When you measure something, it is rarely totally **accurate**.
The accuracy will depend on the units to which you are measuring.

⬤ The highest and lowest values that a measured quantity can be
are called the **upper and lower bounds**.

What are the upper and lower bounds of these heights?

156 cm

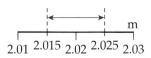

2.02 m

The upper bound is
156.49999.... but this is
really equal to 156.5 so
you write that as the
upper bound.

```
        |←——————→|      cm
155  155.5  156  156.5  157
```

Lower bound 155.5 cm
Upper bound 156.5 cm
155.5 ≤ height (cm) < 156.5

```
        |←———→|      m
2.01  2.015  2.02  2.025  2.03
```

Lower bound 2.015 m
Upper bound 2.025 m
2.015 ≤ height (m) < 2.025

Find the upper and lower bounds for each quantity.
a The population of Keswick is 6000 people (to the nearest 1000).
b The journey from Keswick to Blackburn is 92 miles (to the nearest mile).

When a number
is rounded to the
nearest 1000 then
the number must
lie within ±500
of that amount.

a
```
      |←——————→|         Population
5000  5500  6000  6499  7000      8000
```

b
```
      |←——→|          Miles
91  91.5  92  92.5  93        94
```

Here you can only use whole
number values so the upper
bound is 6499.

The upper bound is 92.5 and
the lower bound is 91.5.

The upper and lower bounds can also be written using inequality signs.

5500 ≤ population ≤ 6499

91.5 miles ≤ journey < 92.5 miles

Exercise 1b

1 Round each of these numbers to

i	1 sf	**ii**	2 sf	**iii**	3 sf

   a 3472   **b** 26 058   **c** 329 154

   d 6 284 903   **e** 294.3   **f** 37.456

   g 4.7654   **h** 0.25459   **i** 0.0059876

   j 0.000 002 6825   **k** 9995

2 Copy and complete this table.

	Level of accuracy	**Lower bound**	**Upper bound**
780	Nearest 10	775	785
2150	Nearest 10		
5600	Nearest 100		
12300	Nearest 100		
47	Nearest whole number		
306	Nearest whole number		

3 Write the upper and lower bounds of these numbers.

   a 1230 (rounded to nearest 10)

   b 2345 (rounded to nearest whole number)

   c 34.5 (rounded to nearest 0.1)

   d 4.56 (rounded to nearest 0.01)

   e 0.567 (rounded to nearest 0.001)

   f 0.067 (rounded to nearest 0.001)

4 Find the upper and lower bounds of these measurements, which are correct to the nearest 5 mm.

   a 35 mm   **b** 40 mm

   c 110 mm   **d** 230 mm

   e 4.5 cm   **f** 10 cm

5 Write your answers to question **4** using inequalities.

Problem solving

6 Identify the upper and lower bounds of these figures.

   a Attendance at a football game was 56 000 (nearest 1000).

   b Speed of sound is 1200 km per hour (nearest 100 km).

   c Beryl's height is 1.6 m (2 sf).

   d Max won about £2 million (1 sf).

   e Width of an ant is 0.093 cm (3 dp).

   f The box contained about 2000 pairs of socks (nearest 100).

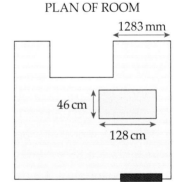

PLAN OF ROOM

1283 mm

46 cm

128 cm

7 Jack buys a cupboard. The cupboard is 128 cm wide and 46 cm deep, both to the nearest cm. The space where Jack wants to put the cupboard is 1283 mm (nearest mm) wide. Explain why the cupboard might not fit.

8 A mini laptop screen is 22.6 cm wide and 16.9 cm high. All measurements are accurate to one decimal place.

   a Calculate the upper and lower bounds for each measurement.

   b Use your answers to calculate the maximum and minimum area of the screen.

   c The width of the screen contains a line of 2560 pixels (nearest 10 pixels). What is the maximum and minimum width of a pixel?

> You can use the upper and lower bounds of quantities in a calculation to place bounds on the answer.

Example

A regular pentagon has sides of length 6 cm.
Find the upper and lower bounds of its perimeter.

A length of 6 cm has been rounded to 1 sf.
The bounds on the side lengths are 5.5 cm and 6.5 cm
The perimeter of a regular pentagon is 5 × side length.
Upper bound   5 × 6.5 = 32.5 cm
Lower bound   5 × 5.5 = 27.5 cm

Sometimes you need to combine a maximum with a minimum amount.

Example

Josiah can run 100 m (to the nearest metre) in 12.4 seconds (3 sf).
Find **a** his maximum possible speed in m/s (metres per second)
  b his minimum possible speed in m/s.

$$\text{Speed} = \frac{\text{Distance}}{\text{Time}}$$

First calculate the upper and lower bounds for each quantity.

Distance ≈ 100 m   Time taken ≈ 12.4 s
Upper bound = 100.5 m   Upper bound = 12.45 s
Lower bound = 99.5 m   Lower bound = 12.35 s

a maximum speed = $\dfrac{\text{maximum distance}}{\text{minimum time}}$ = $\dfrac{100.5}{12.35}$ = 8.14 m/s

b minimum speed = $\dfrac{\text{minimum distance}}{\text{maximum time}}$ = $\dfrac{99.5}{12.45}$ = 7.99 m/s

The maximum speed is when you travel the longest distance in the shortest time.

Example

A lift can hold up to 6 people, with a maximum safe load of
460 kg. A group of 6 people are waiting for the lift. Their
weights are 85 kg, 96 kg, 63 kg, 73 kg, 68 kg and 73 kg
(all measured to the nearest kg). Is it possible for the total
weight of the group to exceed the safe limit?

Add up upper bounds.
85.5 + 96.5 + 63.5 + 73.5 + 68.5 + 73.5 = 461
461 > 460
It is possible for the group to be too heavy for the lift.

Exercise 1c

1 $p = 6$, $q = 3$, $r = 10$

Each number has been rounded to the nearest whole number.

Work out the *smallest* possible value for

a p **b** $p + q$

c $r - p$ **d** qr

e $10r$ **f** $\dfrac{p}{q}$

g $r + q - p$ **h** $\dfrac{pq}{r}$

2 Work out the *largest* possible value for each expression in question **1**.

3 $k = 0.45$, $l = 1.2$, $m = 2.3$ all to 2 sf.

Calculate the difference between the upper and lower bounds of

a k **b** $10k$

c $10m$ **d** $m + l$

e $m - l$ **f** $5k - l$

Problem solving

4 Hercules is carrying out an experiment to see which member of his building team is the best at estimating measurements.

Measurements	Mark's estimate	Barry's estimate	Padraig's estimate	True measurement
Length of room	5 m	520 cm	500 cm	508 cm
Width of room	4 m	430 cm	420 cm	424 cm
Height of room	2.5 m	260 cm	240 cm	248 cm
Area of floor	20 m²	23 m²	21 m²	21.54 m²

a Work out the percentage error of each estimate.

b Who is the best and who is the worst person at estimating? Explain and justify your answers.

Percentage error = $\dfrac{\text{error}}{\text{true measurement}} \times 100$

5 Calculate the maximum and minimum speed in each case.

a Jack runs 1.4 km (1 dp) in 4 minutes (nearest 10 secs)

b A plane flies 3200 miles (2 sf) in 8 hours (nearest hour).

6 a Identify the upper and lower bounds of these quantities.

    i speed of light, 3×10^8 m/s (1 sf)

    ii speed of sound, 1.2×10^3 km/hour (2 sf)

    iii the distance of the Sun from the Earth, 1.47×10^8 km (3 sf)

    iv the distance to Proxima Centauri, the closest star to the Earth, 4×10^{13} km (1 sf).

b Using these measurements estimate the maximum and minimum values for

    i the time it would take light to travel from the Sun to the Earth

    ii the time it would take sound to travel from the Sun to the Earth

    iii the time it would take light and sound to travel from Proxima Centauri to the Earth.

7 $x = 8$, $y = 5$ and $z = 10$ to the nearest whole number.

Work out **i** the smallest **ii** the largest possible value for

a x^3 **b** $\sqrt{y}$ **c** $\dfrac{(x + y)^2}{z^3}$ **d** $x^2 y^3 z^4$

MyMaths.co.uk   Q 1006   **SEARCH**

● You can write a number as the product of its **prime factors** using a method based on repeated division.

Write 700 as the product of its prime factors.

Either use repeated division or a factor tree.

$2\overline{)700}$
$2\overline{)350}$
$5\overline{)175}$
$5\overline{)\,35}$
$\quad\; 7$

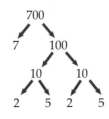

$700 = 2 \times 2 \times 5 \times 5 \times 7$
$\quad\;\;\; = 2^2 \times 5^2 \times 7$

You must divide only by prime numbers. Be systematic and start by testing the smallest prime: 2, 3, 5, 7, 11 ...

● You can find the highest common factor (**HCF**) and the lowest common multiple (**LCM**) of two or more numbers using prime factors and a **Venn diagram**.

Find the HCF and LCM of 56 and 80.

$56 = 2^3 \times 7 \qquad 80 = 2^4 \times 5$

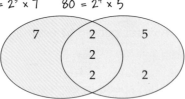

The HCF is the product of the primes in the intersection.

$HCF = 2 \times 2 \times 2$
$\quad\quad\; = 8$

The LCM is the product of the primes in the union.

$LCM = 2 \times 2 \times 2 \times 2 \times 5 \times 7$
$\quad\quad\; = 560$

Put factors that are common to both 56 and 80 in the intersection.

Knowing LCM and HCF allows you to calculate with fractions.

a Work out $\dfrac{3}{5} + \dfrac{1}{4}$ **b** Simplify $\dfrac{36}{84}$

a 20 is the LCM of 4 and 5. **b** 12 is the HCF of 36 and 84.

$\dfrac{3}{5} + \dfrac{1}{4} = \dfrac{12}{20} + \dfrac{5}{20} = \dfrac{17}{20}$

$\dfrac{36}{84} = \dfrac{12 \times 3}{12 \times 7} = \dfrac{3}{7}$

Exercise 1d

1 Write each of these numbers as the product of its prime factors.

a	24	**b**	40	**c**	66
d	84	**e**	108	**f**	140
g	160	**h**	225	**i**	256
j	375	**k**	440	**l**	1440

2 Find the HCF and LCM of these numbers.

a	28 and 42	**b**	35 and 90
c	208 and 273	**d**	350 and 675
e	216 and 576	**f**	1189 and 1827
g	25, 45 and 65	**h**	35, 49 and 63

3 Write these fractions in their simplest form.

a $\dfrac{84}{228}$ **b** $\dfrac{91}{169}$ **c** $\dfrac{350}{434}$

d $\dfrac{315}{525}$ **e** $\dfrac{124}{192}$ **f** $\dfrac{168}{175}$

4 Work out the following, leaving your answers as fractions in their simplest form.

a $\dfrac{5}{8} - \dfrac{1}{7}$ **b** $\dfrac{17}{40} - \dfrac{12}{30}$

c $\dfrac{45}{56} + \dfrac{7}{80}$ **d** $\dfrac{37}{50} + \dfrac{17}{60}$

5 Write the values of x, y and z.

$54 = 2 \times 3^x$

$45 = 5 \times 3^y$

$54 \times 45 = 2 \times 5 \times 3^z$

6 The number 640 can be written as $2^m \times n$, where m and n are prime numbers. Find the values of m and n.

7 For question **2**, parts **a** to **f**, check that HCF × LCM = product of the two numbers.

Problem solving

8 A bus to Thame leaves the bus station every 40 minutes and a bus to Marlow leaves every 30 minutes. At 11:00, two buses leave for Thame and Marlow at the same time.
When is the next time that the buses leave together?

9 A van needs its air filter replaced every 10 000 miles and its oil filter every 6000 miles. After how many miles will they both need replacing at the same time?

10 Two numbers are **co-prime** if they have no common factors (other than 1). By drawing Venn diagrams or otherwise, decide whether each pair of numbers is co-prime.

a	105 and 429	**b**	63 and 715
c	121 and 175	**d**	455 and 693

11 A cuboid is made from 120 small cubes.
The prime factor decomposition of $120 = 2^3 \times 3 \times 5$.
One way of combining the factors is $2^2 \times (2 \times 3) \times 5 = 4 \times 6 \times 5$.
The dimensions of the cuboid could be $4 \times 6 \times 5$.
List the dimensions of three different cuboids that could be made with 120 cubes.

Did you know?

The Ishango bone found near the source of the River Nile is 25000 years old. The marks on it may show the prime numbers.

Check out

You should now be able to ...

Test it ➡

Questions

✓ Round numbers to a given number of significant figures.	7	1, 2
✓ Use rounding to make estimates.	7	3
✓ Find the upper and lower bounds of a calculation or measurement.	7	4 – 8
✓ Use prime factors to find the HCF and LCM of pairs of numbers.	7	9 – 11

Language	Meaning	Example
Significant figures (sf)	The first non-zero figures in a number.	The first two significant figures in 456.7 are 4 (400) and 5 (50).
Rounding	Expressing a number to a given degree of accuracy.	456.7 rounded to 2 sf is 460
Estimate	Use rounding to estimate the answers to calculations.	$\dfrac{3.14 + 2.1}{1.9} \approx \dfrac{3 + 2}{2} \approx \dfrac{5}{2} \approx 2.5$
Upper and lower bounds	The maximum and minimum values that a rounded number or measurement can be.	time = 12.4 s (3 sf) upper bound 12.45 s lower bound 12.35 s or 12.35 s ≤ time < 12.45 s
Prime factors	Prime numbers that are also factors.	$700 = 2^2 \times 5^2 \times 7$
HCF (Highest common factor)	The highest number that is a factor of two or more numbers.	The HCF of 24 and 40 is 8.
LCM (Lowest common multiple)	The lowest number that is a multiple of two or more numbers.	The LCM of 24 and 40 is 120.
Venn diagrams	Overlapping circles used to find HCF and LCM.	Prime factors of 18 Prime factors of 42 3 3 2 7

1 Round 3.0754 to
  a 1 sf   **b** 2 sf   **c** 3 sf

2 Round 0.03597 to
  a 1 sf   **b** 2 sf   **c** 3 sf

3 Estimate the answer to each of these calculations.

  a $\dfrac{62 + 57.1}{18.9}$ **b** $\dfrac{1.9 \times 18.4}{0.77}$

  c $7.6 + 12.5 \times 84$ **d** $\dfrac{41.3 + 6.07}{9.8 - 8.9}$

4 The following measurements were made to the nearest centimetre. Identify the upper and lower bounds of each.
  a 1 cm   **b** 1.5 m
  c 30 mm   **d** 250 mm

5 The following measurements were made to the nearest millimetre. Identify the upper and lower bounds of each.
  a 100 mm   **b** 8 m
  c 11 cm   **d** 1 mm

6 The build cost of an Olympic stadium was £490 million to 2 significant figures. What are the upper and lower bounds of this figure? Use inequality signs in your answer.

7

8 cm

3 cm

The measurements of this rectangle are correct to the nearest mm. Calculate
  a the upper bound of the area
  b the lower bound of the perimeter.

8 A cheetah runs 100 metres (nearest m) in 3.9 s (1 dp). Calculate
  a the lower bound of its speed
  b the upper bound of its speed.
Give your answers in m/s to 3 sf.

9 Write each of these numbers as the product of its prime factors.
  a 84   **b** 126
  c 693   **d** 1430

10 Find the HCF and LCM of these pairs of numbers.
  a 25 and 35   **b** 66 and 234
  c 420 and 315   **d** 1680 and 900

11 Evaluate the following, leaving your answer as a fraction in its simplest form.

  a $\dfrac{13}{60} + \dfrac{11}{105}$ **b** $\dfrac{17}{60} - \dfrac{19}{84}$

What next?

Score		
	0 – 4	Your knowledge of this topic is still developing. To improve look at Formative test: 3C-1; MyMaths: 1005, 1067, 1006, 1034 and 1044
	5 – 9	You are gaining a secure knowledge of this topic. To improve look at InvisiPen: 112, 135, 172, 174 and 183
	10 – 11	You have mastered this topic. Well done, you are ready to progress!

1 Round each of these numbers to

    i   1 sf   **ii**   2 sf   **iii**   3 sf

    a   3729   **b**   34 780   **c**   1.295 64   **d**   163 905

    e   0.370 04   **f**   0.005893   **g**   0.000 285 44   **h**   0.9756

2 Use a calculator to evaluate each expression, giving your answer to three significant figures.

    a 907×301 **b** $21.6 \div 11.9$ **c** 0.0623×17.765 **d** $333 \div 106$

    e $\dfrac{4\pi}{3}$ **f** π^2 **g** $2 \times \left(\dfrac{70}{99}\right)^2$ **h** $49826 - 217$

3 Estimate the answer to each of these calculations by rounding each number to 1 significant figure.

    a $\dfrac{73 \times 2.7}{32.3}$ **b** $\dfrac{261 \times 2.3}{28.1}$ **c** $\dfrac{8.38 \times 5.6}{3.97}$ **d** $\dfrac{4.92 \times 22.5}{0.24}$

    e $\dfrac{243 \times 18.8}{0.2}$ **f** $\dfrac{485 \times 3.25}{32.05}$ **g** $\dfrac{643 + 57.92}{643 - 57.92}$ **h** $\dfrac{147.2 + 102.3}{147.2 - 102.3}$

4 For each of these calculations work out

    i   the exact answer to 2 sf

    ii   the answer if you round intermediate numbers to 2 sf after each step of the calculation.

    a   Toni has 16 boxes of golf balls each containing 24 balls. She paid £179 for the boxes. How much did she pay per golf ball?

    b   Toby the dog eats 375 g of tinned food a day and 750 g of biscuits a week. A 2 kg tin costs £6.99 and a 5 kg bag of biscuits costs £9.49. How much does Toby cost to feed in a year?

5 Write the lower and upper bound of these numbers (all given to 3 sf).

    a   78900   **b**   34500   **c**   4560   **d**   940

    e   789   **f**   234   **g**   123   **h**   61.0

    i   56.9   **j**   45.6   **k**   12.5   **l**   11.5

    m   7.84   **n**   0.123   **o**   0.137   **p**   0.00222

6 A puppy weighs 4 kg to the nearest kg.

    What are the lower and upper bounds for its weight?

    Write your answer using inequalities.

7 The football crowd at Brownfield United was 4400 to the nearest 100.

    The crowd at Blackdown Wanderers was 4360 to the nearest 10.

    Can you be certain which match had the most people watching?

8 For each of the following calculate the maximum and minimum possible
    i perimeter   **ii** area.

    a Football pitch 72 m wide (nearest m) by 104 m long (nearest m).
    b Piece of paper 8 inches wide (nearest inch) by 30 cm long (nearest cm).

9 Work out the largest possible answer to 9.12 − 3.89 if both values have been rounded to 3 sf.

10 If a lawn is 6 m square to the nearest metre, how many square metres of turf must be ordered to be certain of having enough to cover it?

11 A hosepipe is 25 m long to the nearest half metre.
If Harry cuts a piece 4 m long, to the nearest 10 cm, what is the maximum possible length remaining?

---

12 Write these numbers as a product of prime factors.

42   =   ☐ × ☐ × ☐

126   =   ☐ × ☐ × ☐ × ☐

300   =   ☐ × ☐ × ☐ × ☐ × ☐

13 Pair each number with its product.

$2 \times 2 \times 11$	135
$2 \times 11 \times 11$	44
$3^2 \times 5^2$	242
$3^3 \times 5$	273
$3 \times 7 \times 13$	225

14 Find the HCF and LCM of these numbers.
    a 408 and 282   **b** 250 and 75   **c** 85, 119 and 136

15 Put these fractions in ascending order.

$$\frac{3}{8} \qquad \frac{5}{12} \qquad \frac{1}{3} \qquad \frac{5}{6}$$

16 Evaluate the following, leaving your answer as a fraction in its simplest form.
    a $\dfrac{7}{6} + \dfrac{2}{15}$ **b** $\dfrac{9}{20} + \dfrac{8}{35}$ **c** $\dfrac{11}{12} - \dfrac{1}{20}$ **d** $\dfrac{17}{60} - \dfrac{19}{84}$

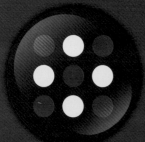

2 Measures, perimeter and area

Introduction

The fastest man in the world is Usain Bolt. In 2009 he ran 100 metres in 9.58 seconds. His average speed for the race was 10.44 m/s but his top speed during the race was 12.27 m/s.

However, Usain would be chasing the tails of creatures such as greyhounds, antelopes, cheetahs and even the domesticated cat, which can all run faster.

What's the point?

Being able to measure distances and times accurately means that you can work out speed, which is highly useful in our increasingly fast-moving world!

Objectives

By the end of this chapter, you will have learned how to ...

- Convert between metric and imperial units.
- Understand whether a formula represents a length, area or volume.
- Calculate the perimeter and area of 2D shapes.
- Understand and use compound measures for speed, density and pressure.

16

Check in

1 Round these numbers to the degree of accuracy given in the brackets.

   a 1434 (nearest 10)   **b** 34.5 (nearest integer)   **c** 36.05 (nearest tenth)

   d 266 (nearest 100)   **e** 6200 (nearest 1000)

2 The radius of a circle is 6 cm.

   Use π = 3.14 to calculate

   a the circumference   **b** the area.

3 A cuboid measures 5 cm by 6 cm by 8 cm.

   Calculate

   a the surface area   **b** the volume.

Starter problem

A ship can sail at a steady speed of 30 km/h. There is enough fuel to last for ten hours. When the ship leaves port there is a strong current of 6 km/h which increases the speed of the ship to 36 km/h. Assuming the captain of the ship intends to use all the fuel, at what distance from port will he have to turn the ship around and head home? Remember on the return journey the strong current will be slowing down the speed of the ship to 24 km/h.

⚫ You usually measure quantities using the **metric** system.

Length

millimetre (mm)
centimetre (cm)
metre (m)
kilometre (km)

1 cm	=	10 mm
1 m	=	100 cm
1 Km	=	1000 m

Area

square centimetre (cm²)
square metre (m²)
hectare (ha)
square kilometre (km²)

1 cm²	=	100 mm²
1 m²	=	10 000 cm²
1 ha	=	10 000 m²
1 Km²	=	1 000 000 m²

Mass

milligram (mg)
gram (g)
kilogram (kg)
tonne (t)

1 g	=	1000 mg
1 Kg	=	1000 g
1 t	=	1000 Kg

Capacity/Volume

millilitre (ml)
centilitre (cl)
litre (ℓ)
cubic centimetre (cm³)
cubic metre (m³)

1 ℓ	=	1000 ml
1 ℓ	=	100 cl
1 ℓ	=	1000 cm³
1 m³	=	1000 litres
1 m³	=	1 000 000 cm³

Example

Convert

a 5 m² to square centimetres

b 60 000 cm³ to litres.

a ×10 000

$1 m^2 = 10 000 cm^2$ $5 m^2 = 5 × 10 000 cm^2$

÷10 000 $= 50 000 cm^2$

b ×1000

1 litre = 1000 cm³   60 000 cm³

÷1000 $= 60 000 ÷ 1000$ litres

   $= 60$ litres

<p.6> When you measure a quantity, the measurement can never be exact. It can be written using upper and lower bounds.

Time is also continuous but you can round time in different ways. Time is not a metric quantity.

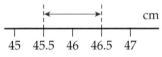

cm

45 45.5 46 46.5 47

46 cm to the nearest cm means a length in the interval 45.5 cm ≤ length < 46.5 cm.

Example

Write these using **inequalities**.

a It takes Majid 11.4 seconds to run 100 m.

b Majid is 15 years old.

Majid is **less than** 16 years old.

a $11.35 ≤ t (sec) < 11.45$

b $15 ≤ t (years) < 16$

Exercise 2a

1 Convert these metric measurements to the units indicated in brackets.

   a 8.5 km (cm)   **b** 80 cm² (mm²)
   c 5000 m² (ha)   **d** 2.5 m³ (cm³)
   e 450 kg (tonnes)   **f** 6500 cm³ (ℓ)
   g 1 million mm (km)   **h** 6000 cm² (m²)
   i 0.6 m² (cm²)   **j** 435 ℓ (m³)
   k 0.008 m² (cm²)   **l** 123 000 m² (ha)
   m 6.5 million g (tonnes)

2 Each of these measurements has been given to a particular accuracy, shown in brackets. Write each measurement as an inequality.

   a 78 m (nearest m)
   b 4 tonnes (nearest t)
   c 3.2 kg (nearest 0.1 kg)
   d 36.0 g (nearest 0.1 g)
   e 5.24 km (nearest 0.01 km)

Problem solving

3 The Westfield Shopping Centre in London has a retail floor space of 150 000 m².
   A football pitch has an area of 0.6 hectares.
   How many football pitches would cover the floor space?

4 Use these metric-imperial equivalents to convert these measurements to the units indicated in brackets.

Length	Capacity	Mass
1 inch ≈ 2.5 cm	1 gallon ≈ 4.5 litres	1 oz (ounce) ≈ 30 g
1 yard ≈ 1 metre	1 pint ≈ 0.6 litres	1 kg ≈ 2.2 lb (pounds)
5 miles ≈ 8 km	1 pint ≈ 600 mℓ	

   a 8.5 gallons (litres)   **b** 3.5 pints (litres)
   c 30 kg (lb)   **d** 6 oz (g)
   e 22.5 feet (m)   **f** 123.2 lb (kg)
   g 150 ml (pints)   **h** 47.5 miles (km)

5 Pete is 15 and Amy is 14 years old. Work out the maximum number of days that Pete is older than Amy.

6 Greengrocers used to weigh vegetables on balance scales using a set of weights.

   a Show that you can obtain every mass from 1 kg to 15 kg using a set of weights containing 1 kg, 2 kg, 4 kg and 8 kg weights.
   b Show that you can obtain every mass from 1 kg to 40 kg using a set of weights containing 1 kg, 3 kg, 9 kg and 27 kg weights.

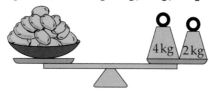

6 kg of potatoes

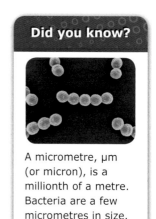

Did you know?

A micrometre, μm (or micron), is a millionth of a metre. Bacteria are a few micrometres in size.

 MyMaths.co.uk

You measure a distance in units of **length**, for example, centimetres.

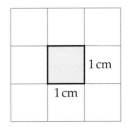

You measure an **area** in square units.
This is 1 square centimetre (1 cm²).

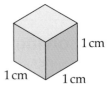

1 cm

1 cm

You measure a **volume** in cubic units.
This is 1 cubic centimetre (1 cm³).

1 cm

1 cm   1 cm

- A length has one **dimension**: length.
- An area has two dimensions: length × length.
- A volume has three dimensions: length × length × length.
- Numbers have no dimension.

$\frac{1}{2}, \frac{1}{3}, \frac{4}{3}, 3, \pi, 4$
are examples of numbers with no dimensions.

Example

Decide whether each quantity is a length, an area, a volume or a number.

a 3.141592
b The height of a building
c The surface of a jigsaw piece

a Number
b Length
c Area

The clue is in the language – surface relates to area.

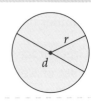

You can use dimensions to decide if a **formula** measures length, area or volume.

Example

Ameer has to choose a formula for the circumference of a circle.
These are his choices.
$C = \pi r^2$ $C = \pi$ $C = 2\pi r$ $C = \pi r^2 d$

Which of these formulae could represent the circumference of a circle?

Circumference is a length.

πr^2	No	number × length × length	= area
π	No	number	= number
$2\pi r$	Yes	number × number × length	= length
$\pi r^2 d$	No	number × length × length × length = volume	

Exercise 2b

1 Decide whether each quantity is a length, an area, a volume or a number.

   a The space inside a car

   b $10\,cm^2$

   c The distance to your home

   d The surface of a carpet

   e The perimeter of a cricket pitch

   f π

   g The surface of a desk

   h Your height

   i The perimeter of a rectangle

   j $9\,mm$

   k The depth of a lake

   l The space inside a box

   m The height of a wall

   n 3

   o The surface of a 3D shape

   p $8\,m^3$

1 **q** The distance from London to Oxford

   r The width of a car

   s The height of a mountain

   t The capacity of a bottle

2 State whether these expressions represent a length, an area, a volume or a number.

   a length × width × height

   b length × width

   c π × diameter

   d number × length

   e π × radius × radius

   f length + width + length + width

   g $2 \times \pi \times radius$

   h base × perpendicular height

   i π × radius × radius × height

   j $\frac{1}{2}$ × base × perpendicular height

   k π × radius × slant height

Problem solving

3 The formulae for the surface area and the volume of a sphere, with radius r, are given by

   surface area $= 4\pi r^2$ volume $= \frac{4}{3}\pi r^3$.

   Show that the dimensions of each formula are correct.

4 Here is a sector of a circle of radius r.

   Which expression might represent

   a the arc length

   b the area?

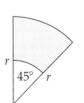

$$\frac{1}{8}\pi r^2 \qquad \frac{1}{4}\pi r \qquad 45r \qquad \frac{1}{8}r^3 \qquad \frac{\pi}{4}$$

Did you know?

$\pi = 3.1415926535...$
9502884197169399...
0781640628620899...
0821480865132823...
9821172553594081...
...3141105923344...

In 1873, after 15 years calculating by hand, William Shanks published the value of π to 707 decimal places. Only in 1944 were mistakes found starting at the 527th digit!

5 A cuboid has length ℓ, width w and height h.

   Find a formula in terms of ℓ, w and h for

   a the total length of all the edges

   b the surface area

   c the volume of the cuboid.

   Check that each formula has the correct dimensions.

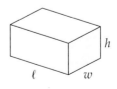

 MyMaths.co.uk   Q 1096   **SEARCH**

You can calculate the **perimeter** and area of a 2D shape using these formulae.

> ● Perimeter of a **rectangle**
> $= \ell + w + \ell + w = 2(\ell + w)$
> ● Area of a rectangle $= \ell \times w$

> ● **Circumference** of a **circle** $= \pi d = 2\pi r$
> ● Area of a circle $= \pi \times r \times r = \pi r^2$

where d = **diameter**

r = **radius**

and $\pi = 3.141592....$

$d = 2r$

> ● Area of a **triangle**
> $= \dfrac{1}{2} \times b \times h$

> ● Area of a **parallelogram**
> $= b \times h$

> ● Area of a **trapezium**
> $= \dfrac{1}{2} \times (a + b) \times h$

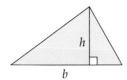

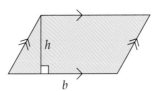

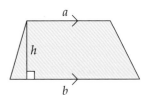

Calculate the **surface area** of this triangular **prism**.

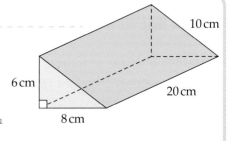

Area of two triangles $= \frac{1}{2} \times 8 \times 6 + \frac{1}{2} \times 8 \times 6$

$\qquad\qquad\qquad = 24\,cm^2 + 24\,cm^2$

Area of three rectangles $= 10 \times 20 + 6 \times 20 + 8 \times 20$

$\qquad\qquad\qquad\quad = 200\,cm^2 + 120\,cm^2 + 160\,cm^2$

Total surface area $= 24\,cm^2 + 24\,cm^2 + 200\,cm^2 + 120\,cm^2 + 160\,cm^2$

$\qquad\qquad\qquad = 528\,cm^2$

Calculate the circumference and area of this circle.

Give your answers (i) in terms of π and (ii) to 3 sf.

Circumference $= 2 \times \pi \times$ radius

$\qquad\qquad\quad = 2 \times \pi \times 4 = 8\pi\,cm = 25.1\,cm \;(3\,sf)$

Area $= \pi \times$ radius^2

$\qquad = \pi \times 4^2 = 16\pi\,cm^2 = 50.3\,cm^2 \;(3\,sf)$

Answers given in terms of π are exact.

Exercise 2c

1 Calculate the perimeter and area of these shapes.

Use π = 3.14 if required.

a
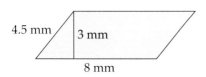
8 cm
3 cm

b
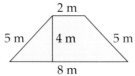
5 m
5 m

c

10 cm   10 cm
6 cm
16 cm

d
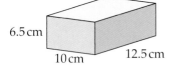
4.5 mm   3 mm
8 mm

e
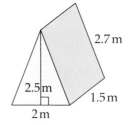
2 m
5 m   4 m   5 m
8 m

f

6 cm

2 Calculate the surface area of each of these prisms.
State the units of each answer.

a
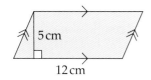
6.5 cm
10 cm   12.5 cm

b

2.7 m
2.5 m
2 m   1.5 m

3 In terms of π and r, calculate the area of
  a the small square
  b the circle
  c the large square.

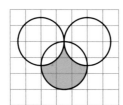
r

Did you know?

The word 'trapezium' originates from the shape made by the ropes and bar of an old-fashioned flying trapeze.

Problem solving

4 The areas of the parallelogram and the trapezium are the same.
What is the value of a + b?

5 cm
12 cm

a
4 cm
b

5 Three circles, each of diameter 3 cm, are drawn on 1 cm squared paper.
Calculate the perimeter and area of the shaded shape.

Speed measures how fast something moves, or how quickly distance changes 'per unit time'. If the speed is constant

● Speed (S) = $\dfrac{\text{Distance travelled (D)}}{\text{Time taken (T)}}$

Speed is measured in metres per second (m/s), miles per hour (mph) or kilometres per hour (km/h).

If the speed is changing then the same formula gives the **average speed**.

Example

A cyclist travels 36 km in 1 hour 15 minutes.
Calculate the average speed in
a   kilometres per hour
b   metres per second.

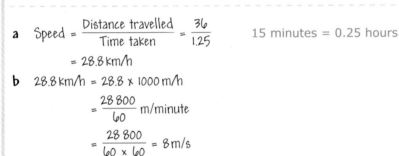

a   Speed = $\dfrac{\text{Distance travelled}}{\text{Time taken}} = \dfrac{36}{1.25}$ 15 minutes = 0.25 hours

= 28.8 km/h

b   28.8 km/h = 28.8 × 1000 m/h

= $\dfrac{28\,800}{60}$ m/minute

= $\dfrac{28\,800}{60 \times 60}$ = 8 m/s

You can use a formula triangle to rearrange the formulae for speed, density and pressure.

$D = S \times T$

$S = \dfrac{D}{T}$ $T = \dfrac{D}{S}$

Speed is an example of a **compound measure**.
A compound measure uses a combination of measurements and units.
Density measures how heavy something is 'per unit volume'.

● Density (D) = $\dfrac{\text{Mass (M)}}{\text{Volume (V)}}$

Density is measured in grams per cubic centimetre (g/cm³) or kilograms per cubic metre (kg/m³).

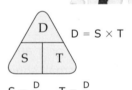

$M = D \times V$

$D = \dfrac{M}{V}$ $V = \dfrac{M}{D}$

Pressure measures how force acts 'per unit area'.

● Pressure (P) = $\dfrac{\text{Force (F)}}{\text{Area (A)}}$

Pressure is measured in newtons per square metre (N/m²).

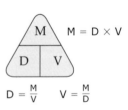

$F = P \times A$

$P = \dfrac{F}{A}$ $A = \dfrac{F}{P}$

A newton (N) is a unit of force named after Isaac Newton.

Exercise 2d

1   Calculate the average speed
    when you travel
    a   100 km in 2 hours
    b   350 km in 5 hours
    c   80 miles in 2 hours
    d   30 km in 30 minutes
    e   250 m in 5 s
    f   240 m in 3 minutes.

2   Calculate the distance travelled
    when you travel at
    a   25 km/h for 3 hours
    b   70 km/h for 2 hours

2   c   50 km/h for 3.5 hours
    d   40 m/s for 15 seconds
    e   55 m/s for 10 seconds
    f   30 m/s for 2 minutes.

3   Kathy ran 100 metres in
    25 seconds.
    Jayne ran 400 metres in
    125 seconds.
    a   Calculate their speeds in
        metres per second.
    b   Who ran faster, Kathy or
        Jayne?

Did you know?

When an aircraft
reaches the speed
of sound, a conical
pressure wave forms
and there is a sonic
boom.

Problem solving

4   The safety valve on a boiler releases the pressure if the pressure is
    greater than 4 million newtons per square metre.
    If the force is 1 500 000 newtons and the area is 0.5 square metres,
    will the safety valve be activated?

5   In 2002, the fastest ever lap of a Formula One racing track was
    recorded at Monza in Italy. Juan Pablo Montoya completed one lap
    of 5783 metres in 80 seconds. Calculate his average speed in
    a   metres per second   b   kilometres per hour.

6   In 1997 Andy Green, an RAF fighter pilot, set the World Land
    Speed Record at Black Rock Desert in Nevada, USA, when he
    travelled at 768 mph in the ThrustSSC.
    a   Calculate his average speed in kilometres per hour.
    At sea level the speed of sound is 1225 km/h.
    b   Did Andy Green travel faster than the speed of sound?

7   The density of cork is 0.25 g/cm^3 and the density of water is
    1.0 g/cm^3. Calculate the mass of
    a   a 10 cm cube of cork   b   a litre of water.

8   Many singers use the words 'a million miles' in their lyrics.
    'It feels like you're a million miles away.' (*Rihanna*)
    'I can't stay a million miles away.'   (*Offspring*)
    'I'm still a million miles from you.'   (*Bob Dylan*)
    How many years would it take you to walk a million miles at a
    speed of 3 miles per hour?

Check out

You should now be able to ...

✓ Convert between metric and imperial units.	7	1, 2
✓ Understand whether a formula represents a length, area or volume.	8	3
✓ Calculate the perimeter and area of 2D shapes.	8	4 – 7
✓ Understand and use compound measures for speed, density and pressure.	7	8, 9

Language	Meaning	Example
Dimension	A measurement of length.	A area has 2 dimensions (2D) length × length
Volume	A 3D measure of space, measured in cubic units.	A cuboid 4 cm × 3 cm × 2 cm has a volume of 24 cm³
Compound measure	A unit with more than one measure.	The car travelled at 56 km per hour
Speed	The rate of change of distance. $$\text{speed} = \frac{\text{distance travelled}}{\text{time taken}}$$	The bullet travelled at 200 m/s (metres per second)
Density	Measures the mass of a substance per unit volume.	The density of water is 1000 kg/m³
Pressure	Measures the force on an object per unit area.	The pressure under an elephant's foot is 80 000 Newtons/m²

1 Convert between these metric measurements.
  a 6.82 m to mm **b** 14 cl to litres
  c 0.82 kg to g **d** 0.7 cm² to mm²
  e 8 m² to cm² **f** 12 l in cm³

2 Convert these imperial measurements to metric.
  a 8 pints to litres **b** 22 lb to kg
  c 120 miles to km **d** 25 inches to m

3 State whether each expression represents a length, an area or a volume.
  a height × width ÷ 2
  b $\frac{1}{4}$ × (diameter)² × length
  c radius × π
  d $\frac{1}{2}$ × base × height + width × length

4 Calculate the surface area of the prism.

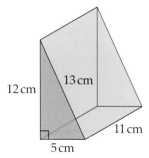

12 cm 13 cm 11 cm 5 cm

5 The parallelogram and the trapezium have the same area. What is the height of the parallelogram?

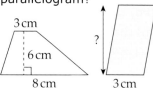

3 cm 6 cm 8 cm ? 3 cm

6 Calculate
  a the area
  b the circumference
  of the circle.
  Give your answers to 1 dp.

19 cm

7 The diagram shows a quarter of a circle with radius 12 cm.
  Calculate
  a the area
  b the perimeter.
  Give your answers in terms of π.

12 cm

8 A 5 mm cube diamond weighs 0.4 g. Calculate the density of the diamond in g/cm³.

9 A car travels at 85 km/h.
  a How far does it travel in 42 minutes?
  b How long does it take to travel 17 km?

What next?

Score		
	1 – 4	Your knowledge of this topic is still developing. To improve look at Formative test: 3C-2; MyMaths: 1096, 1108, 1129, 1088, 1083, 1121 and 1246
	5 – 7	You are gaining a secure knowledge of this topic. To improve look at InvisiPen: 315, 327, 333, 334, 335 and 353
	8 – 9	You have mastered this topic. Well done, you are ready to progress!

2a

1 Convert these metric measurements to the units indicated in brackets.

Use π = 3.14 if required

   a 7.5t (kg)   **b** 480mm (m)   **c** $1.5\,cm^2$ (mm^2)   **d** $15000\,cm^2$ (m^2)

   e $3500\,m^2$ (ha)   **f** $4.8\,m^3$ (cm^3)   **g** 19.6km (cm)   **h** 800000mg (kg)

   i $0.5\,m^2$ (cm^2)   **j** 1.8t (kg)   **k** 0.3ml (litres)   **l** 7ha (m^2)

2 Convert these measurements to the units indicated in brackets.
Use the conversions in Exercise **2a** question **4**.

   a 9oz (g)   **b** 3 pints (ml)   **c** 60ft(m)   **d** 6 gallons (litres)

   e 181 (pints)   **f** 45kg (lb)   **g** 1.75m (in)   **h** 36in (cm)

2b

3 **a** Which formula could give the curved surface area of a cone? Explain your reasoning.

   $A = πl$ $A = πrl$ $A = πr^2l$ $A = πr$ $A = π$

   b Explain why the formula for the curved surface area of a cone is unlikely to be $A = rl$.

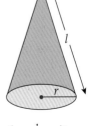

4 The plan view of a sports arena is shown.
Which expression could be correct for

   a the perimeter   **b** the area?

   $2l + 2πr$ $2rl + πr^2$ $πr^2l$ $2π$

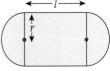

2c

5 For each shape calculate its

   i perimeter   **ii** area.   **a**

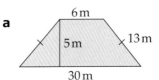

   b

6 Calculate the total surface area of each prism.

   a   **b**

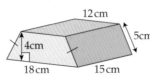

7 Four circles of radius 6cm are placed touching each other.
The centres of the circles are joined to form a square.
Calculate the perimeter and the area of the shaded shape.

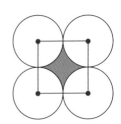

8 Copy and complete the table to show the distances, times and speeds for these journeys.

	Distance travelled (km)	Time taken	Speed (km/h)
a	112	5 hours	
b	84	3 hours	
c	15	30 mins	
d	36	4 hours 30 mins	
e		6 hours	25
f		5 hours 30 mins	6
g		1 hour 45 mins	52
h	450		20
i	81		18
j	19		4

9 The speed limits for vehicles in Ireland are

Motorways	120 km/h
National roads	100 km/h
Other rural roads	80 km/h
Built-up areas	50 km/h

Convert these speeds to miles per hour.

10 a A 2 centimetre cube of lead has a mass of 90.7 grams.
Calculate the density of lead in g/cm^3.

b A bar of gold forms a cuboid 12 cm by 5 cm by 7.5 mm.
The density of gold is 19.3 g/cm^3.
How heavy is the bar of gold?

C An African elephant weighs 2.875 tonnes.

1 kg exterts a force of 10 N

The average diameter of the elephant's feet is 22 cm.
Calculate the pressure, in N/cm^2, underneath the elephant's feet?

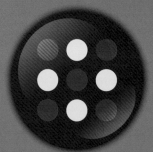

3 Expressions and formulae

Introduction

The most famous scientists have used mathematical formulae to describe their ideas. When an apple famously dropped on his head, Sir Isaac Newton used the formula

$$F = \frac{Gm_1m_2}{r^2}$$

to describe what he called gravity.

When Albert Einstein said that you could convert between mass and energy he changed the world. He used the formula

$$E = mc^2$$

to show how it could be calculated.

What's the point?

A formula shows how one quantity is linked to another. Using formulae helps us to understand how our world is connected.

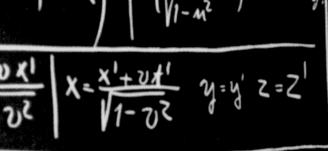

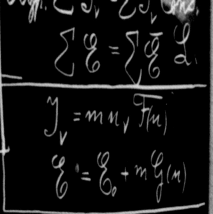

Objectives

By the end of this chapter, you will have learned how to …

- ● Know and use the index laws.
- ● Multiply brackets in two linear expressions.
- ● Factorise expressions by taking common factors.
- ● Derive simple identities, including expansion of two linear brackets.
- ● Derive formulae and substitute values in formulae to find unknown variables.
- ● Change the subject of a formula.

Check in

1 Simplify these algebraic expressions, where possible.

   a $x + x + x$ **b** $6y + 3y$ **c** $12a - 5a$

   d $5b - 2$ **e** $p + 5p - 2p$ **f** $4a + 5b - 3a$

   g $x^2 - 3x + 5x$ **h** $10y - y^2 + 2y^2 - 4y$

2 The formula to find the perimeter, P, of a rectangle with length, l, and width, w, is $P = 2(l + w)$. Find P when

   a $l = 12\,cm$ and $w = 9\,cm$ **b** $l = 16\,cm$ and $w = 7\,cm$.

3 Expand these brackets.

   a $2(x + 5)$ **b** $7(p - 3)$ **c** $5(2k + 1)$ **d** $3(4 - y)$

   e $x(x + 2)$ **f** $a(b - 6)$ **g** $4p(p + q)$ **h** $3t(5 - 2t)$

Starter problem

This square grid is numbered from 1 to 100.

A 2×2 square is shaded in as shown on the grid.

The numbers in the opposite corners of the

2×2 square are multiplied together.

Investigate.

1	2	3	4	5	6	7	8	9	10
11	12	13	14	15	16	17	18	19	20
21	22	23	24	25	26	27	28	29	30
31	32	33	34	35	36	37	38	39	40
41	42	43	44	45	46	47	48	49	50
51	52	53	54	55	56	57	58	59	60
61	62	63	64	65	66	67	68	69	70
71	72	73	74	75	76	77	78	79	80
81	82	83	84	85	86	87	88	89	90
91	92	93	94	95	96	97	98	99	100

3a Index laws 1

Repeated multiplications of the same number can be written in **index** form. This often makes it quicker and clearer.

$5 \times 5 \times 5 = 5^3$ has **base** 5 and index or **power** 3.

$n \times n \times n \times n \times n \times n \times n = n^7$ has base n and index 7.

Example

Evaluate

a $3^2 \times 3^4$ **b** $5^6 \div 5^4$ **c** $(4^2)^3$

---

a $3^2 \times 3^4$
$= (3 \times 3) \times (3 \times 3 \times 3 \times 3)$
$= 3^6$

Add the indices
$2 + 4 = 6$

b $5^6 \div 5^4$
$= \dfrac{5 \times 5 \times 5 \times 5 \times 5 \times 5}{5 \times 5 \times 5 \times 5}$
$= 5^2$

Subtract the indices
$6 - 4 = 2$

c $(4^2)^3$
$= (4^2) \times (4^2) \times (4^2)$
$= (4 \times 4) \times (4 \times 4) \times (4 \times 4)$
$= 4^6$

Multiply the indices
$2 \times 3 = 6$

● The index laws state

$p^a \times p^b = p^{a+b}$ $p^a \div p^b = p^{a-b}$ $(p^a)^b = p^{ab}$

You can use the index laws only when the base numbers are the same.

Example

Simplify

a $2^6 \times 2^3$ **b** $7^8 \div 7^3$ **c** $(6^3)^4$

     Add   Subtract   Multiply

a $2^6 \times 2^3 = 2^{6+3}$ **b** $7^8 \div 7^3 = 7^{8-3}$ **c** $(6^3)^4 = 6^{3 \times 4}$
        $= 2^9$ $= 7^5$ $= 6^{12}$

● **Indices** in algebra follow the same rules as arithmetic.

Example

Simplify

a $2x^3 \times 5x^2$ **b** $12y^7 \div 4y^3$ **c** $4a^3 \times 2b^2$

Treat numbers and symbols seperately.

a $2x^3 \times 5x^2$
    $= 2 \times 5 \times x^{3+2}$
    $= 10x^5$

b $12y^7 \div 4y^3$
    $= (12 \div 4) \times y^{7-3}$
    $= 3y^4$

c $4a^3 \times 2b^2$
    $= 8a^3b^2$

The index laws do not apply in part **c** as a and b are different base numbers. Simplify using the rules of algebra.

7

Exercise 3a

1 Evaluate these expressions without using a calculator.

  a 5^2 **b** 2^3

  c 10^4 **d** 3^4

  e 2^7 **f** 1^{12}

  g $(-1)^5$ **h** $(-4)^3$

  i $(-3)^4$ **j** $(-5)^2$

  k $(-10)^3$ **l** $(-6)^2$

  m $\left(\dfrac{1}{2}\right)^2$ **n** 0^3

  o $\left(-\dfrac{1}{3}\right)^2$ **p** $\left(-\dfrac{1}{4}\right)^3$

  q $(-1)^{101}$ **r** $(-1)^{1001}$

2 Simplify these expressions, leaving your answers in index form.

  a $3^2 \times 3^5$ **b** $4^6 \times 4^3$

  c $8^{12} \div 8^5$ **d** $(2^3)^4$

  e $x^{10} \div x^4$ **f** $a^3 \times a^6$

  g $(y^2)^9$ **h** $\dfrac{n^9}{n^7}$

3 Simplify these expressions, leaving your answers in index form.

  a $3a^2 \times 2a^5$ **b** $10b^6 \div 5b^4$

  c $(2x^4)^2$ **d** $\dfrac{12y^5}{3y^4}$

  e $5p^4 \times 3p^{10}$ **f** $(3q^7)^4$

  g $2m^3 \times 3m^2 \times 4m$ **h** $\dfrac{4n^{10} \times 3n^5}{6n^3}$

Problem solving

4 Write a simplified expression for the area of each shape.

a

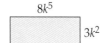

$8k^5$ $3k^2$

b

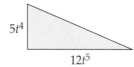

$5t^4$ $12t^5$

c

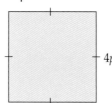

$4p^7$

d

$10w^3$

5 Copy and complete each of these statements.

  a $3^3 \times \square = 3^{10}$ **b** $\square \div 5^9 = 5^4$ **c** $(x^{\square})^3 = x^{12}$

  d $\square\, t^{10} \div 3t^{\square} = 2t^3$ **e** $5k^{\square} \times \square\, k^6 = 20k^8$ **f** $(3a^5)^{\square} = 81a^{\square}$

6 Are these statements true or false? Give reasons for each answer.

  a $a^5 + a^2$ simplifies to a^7

  b $2x^3 \times 5y^7$ simplifies to $10x^3y^7$

  c $2^a \times 2^b$ simplifies to 4^{a+b}

7 Solve these equations.

  a $x^2 = 36$ **b** $x^{10} = 1$

  c $2^x = 64$ **d** $3^x = 27$

  e $x^3 = -125$ **f** $x^5 = -32$

  g $4^{x-1} = 16$ **h** $2^{x-2} = 128$

  i $3^{5-x} = 81$ **j** $5^{4-x} = 25$

  k $x^2 = 3^4$ **l** $2^6 = x^3$

3b Index laws 2

Powers don't always have to be positive whole numbers.

Example

Simplify $x^4 \div x^4$ and comment on your result.

Using the law of indices.

$x^4 \div x^4 = x^{4-4} = x^0$

When you divide you subtract indices.

Using the laws of arithmetic.

$x^4 \div x^4 = 1$

Any non-zero number divided by itself $= 1$.

x to the power zero equals one, $x^0 = 1$.

⬤ For zero powers, $x^0 = 1$ for all non-zero values of x

For example, $5^0 = 1$, $1234^0 = 1$ and $(-10)^0 = 1$

Example

Simplify $2^3 \div 2^6$ and comment on your result.

Using the law of indices.

$2^3 \div 2^6 = 2^{3-6} = 2^{-3}$

Using the laws of arithmetic.

$\dfrac{2^3}{2^6} = \dfrac{\cancel{2} \times \cancel{2} \times \cancel{2}}{\cancel{2} \times \cancel{2} \times \cancel{2} \times 2 \times 2 \times 2} = \dfrac{1}{2^3}$

Two to the power minus three equals one over two to the power three, $2^{-3} = \dfrac{1}{2^3}$

⬤ For negative powers, $x^{-n} = \dfrac{1}{x^n}$

For a negative index, write the **reciprocal** with a positive index.

Example

Work out these expressions and comment on your results.

a $3^{\frac{1}{2}} \times 3^{\frac{1}{2}}$

b $5^{\frac{1}{3}} \times 5^{\frac{1}{3}} \times 5^{\frac{1}{3}}$

a Using the law of indices.

$3^{\frac{1}{2}} \times 3^{\frac{1}{2}} = 3$

$\sqrt{3} \times \sqrt{3} = 3$ by definition

$3^{\frac{1}{2}} = \sqrt{3}$

b Using the law of indices.

$5^{\frac{1}{3}} \times 5^{\frac{1}{3}} \times 5^{\frac{1}{3}} = 5$

$\sqrt[3]{5} \times \sqrt[3]{5} \times \sqrt[3]{5} = 5$ by definition

$5^{\frac{1}{3}} = \sqrt[3]{5}$

⬤ For **fractional powers**, $x^{\frac{1}{n}} = \sqrt[n]{x}$

The nth root, $\sqrt[n]{}$, is that number which when multiplied by itself n times, gives the original value.

Example

Evaluate

a 10^0

b 3^{-2}

c $32^{\frac{1}{5}}$

d $(x^3)^{-5}$

a $10^0 = 1$

b $3^{-2} = \dfrac{1}{3^2}$

$= \dfrac{1}{9}$

c $32^{\frac{1}{5}} = \sqrt[5]{32}$

$= 2$

$2 \times 2 \times 2 \times 2 \times 2 = 32$

d $(x^3)^{-5} = x^{-15}$

$= \dfrac{1}{x^{15}}$

Exercise 3b

1 Evaluate these expressions without using a calculator.

   a 4^0 **b** $9^{\frac{1}{2}}$

   c 7^{-1} **d** $8^{\frac{1}{3}}$

   e 5^{-2} **f** 100^0

   g $49^{\frac{1}{2}}$ **h** 4^{-3}

   i $27^{\frac{1}{3}}$ **j** $27^{-\frac{1}{3}}$

2 Simplify these expressions. Write your answer in the form $\frac{1}{p^n}$ where appropriate.

   a $2^8 \times 2^{-3}$ **b** $5^2 \times 5^{-4}$

   c $3^2 \div 3^5$ **d** $\dfrac{8^3}{8^7}$

   e $(2^4)^{-3}$ **f** $a^3 \times a^{-3}$

   g $(x^{-5})^2$ **h** $\dfrac{n^3}{n^{-4}}$

3 Simplify these expressions, leaving your answer in index form.

   a $3^{\frac{1}{2}} \times 3^{\frac{1}{2}}$ **b** $(8^{\frac{1}{3}})^2$

   c $10 \div 10^{\frac{1}{4}}$ **d** $4^{\frac{1}{3}} \times 4^{-1\frac{1}{3}}$

   e $(x^2)^{\frac{1}{5}}$ **f** $y^{\frac{5}{4}} \div y$

   g $(a^3)^{\frac{1}{2}} \times a^{\frac{1}{2}}$ **h** $\dfrac{b^{-\frac{1}{4}} \times b^2}{b^{\frac{3}{4}}}$

Problem solving

4 Solve these equations for x.

   a $x^{\frac{1}{2}} = 4$ **b** $x^{\frac{1}{3}} = 2$ **c** $2x^{\frac{1}{2}} = 10$ **d** $5x^{\frac{1}{3}} = 20$

   e $x^{-2} = \dfrac{1}{4}$ **f** $x^{-3} = 8$ **g** $18x^{-2} = 2$ **h** $81x^{-3} = 3$

5 A group of friends write these cards.

   Isla   Grace   Flora   Isobel   Jason

   a Simplify each expression.

   b Arrange the friends so that the values of their expressions are in ascending order if

       i $x = 4$ **ii** $x = \dfrac{1}{4}$

6 Given that $a = 4$ and $b = 8$ find the value of these expressions.

   a $a^{\frac{1}{2}}$ **b** b^{-2} **c** $2a^{-3}$ **d** $a^2b^{\frac{1}{3}}$ **e** $ab^{-1} + 1$

7 Write simplified expressions for the areas of these shapes.

   a   **b**

$3k^{-3}$ $5t^{\frac{1}{2}}$ $2t$

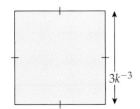 **MyMaths**.co.uk   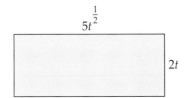 Q 1045   **SEARCH**

● To **expand** a bracket multiply each term inside the bracket by the term outside the bracket.

4 lots of $x + 7 = 4(x+7)$
$\qquad\qquad\qquad = 4x + 28$

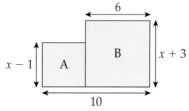

Example

Write an expression for the total area of this shape.
Simplify your answer.

Area of rectangle A $= 4(x - 1)$
Area of rectangle B $= 6(x + 3)$

Total area of shape $= 4(x - 1) + 6(x + 3)$
$\qquad\qquad\qquad\quad = 4x - 4 + 6x + 18$
$\qquad\qquad\qquad\quad = 10x + 14$

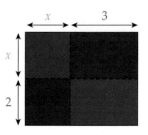

Length of A $= 10 - 6 = 4$

Look at this large rectangle.
Area of rectangle = length × width
$\qquad\qquad\qquad = (x + 3)(x + 2)$
Now look at the four small rectangles.
Areas of rectangles $= x \times x + 3 \times x + x \times 2 + 3 \times 2$
$\qquad\qquad\qquad\qquad = x^2 + 3x + 2x + 6$
$\qquad\qquad\qquad\qquad = x^2 + 5x + 6$
The areas are the same so $(x + 3)(x + 2) = x^2 + 5x + 6$

This is a **quadratic** expression as the highest power of x is x^2.

● To expand **double brackets,** multiply each term in the first bracket by each term in the second bracket.

Example

Expand these double brackets and simplify your answer.
a $(x + 3)(x + 5)$ **b** $(x + 2)(x - 3)$ **c** $(x - 4)^2$

a $(x + 3)(x + 5)$
$\quad = x^2 + 3x + 5x + 15$
$\quad = x^2 + 8x + 15$

b $(x + 2)(x - 3)$
$\quad = x^2 + 2x - 3x - 6$
$\quad = x^2 - x - 6$

c $(x - 4)^2$
$\quad = (x - 4)(x - 4)$
$\quad = x^2 - 4x - 4x + 16$
$\quad = x^2 - 8x + 16$

To help you remember, draw a 'smiley face'!
$(x + 3)(x + 2)$

Exercise 3c

1 Use the diagrams to help expand these double brackets.

a $(x + 5)(x + 1)$

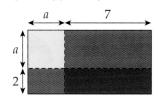

b $(a + 2)(a + 7)$

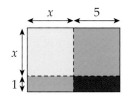

2 Expand these double brackets and simplify your answers.

a $(x + 1)(x + 3)$ **b** $(a + 3)(a + 4)$
c $(k + 2)(k + 9)$ **d** $(t + 6)(t + 3)$
e $(p + 5)^2$ **f** $(n + 10)^2$

3 Expand these double brackets and simplify your answers.

a $(y + 4)(y - 3)$ **b** $(b - 2)(b + 5)$
c $(d + 3)(d - 5)$ **d** $(g + 2)(g - 4)$
e $(q + 3)(q - 3)$ **f** $(m + 5)(m - 5)$

4 Expand these double brackets and simplify your answers.

a $(a - 1)(a - 2)$ **b** $(t - 5)(t - 2)$
c $(x - 3)(x - 8)$ **d** $(m - 4)(m - 7)$
e $(n - 5)^2$ **f** $(p - 7)^2$

5 Expand the brackets and simplify these expressions.

a $(x + 2)(x + 1) + (x + 5)(x + 3)$
b $(y + 2)^2 + (y + 2)(y - 1)$
c $(a - 3)(a + 8) + (a - 4)(a + 2)$
d $(b - 7)(b + 2) + (b - 3)^2$
e $(p - 5)(p + 6) - (p + 2)(p + 5)$
f $(q - 7)^2 - (q + 6)(q - 6)$

Problem solving

6 Write an algebraic expression for the area of each shape.
Expand the brackets and simplify your answer.

a
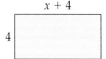

b
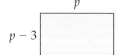

c
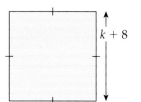

d
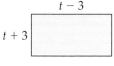

7 **a** Show that $(n + 1)^2 \cdot n^2 + 2n + 1$

b Use this result to calculate
  i 11^2 **ii** 51^2 **iii** 101^2

c Write and expand a similar linear expression in order to calculate
  i 19^2 **ii** 49^2 **iii** 99^2

8 Expand the brackets and simplify these expressions.

a $(2x + 3)(x + 4)$
b $(3x - 4)(2x + 1)$
c $(2x + 1)(x + 3) + (x + 1)(2x + 5)$
d $(3x - 2)(x + 4) + (2x - 1)(2x + 3)$
e $(x + y)(2x + 3y)$
f $(x + y)^2 - (x - y)^2$
g $(1 + x)^3$
h $(x + y)^3$

3d Factorising expressions

Factorising is the reverse of expanding.

> ⬤ To **factorise** an expression you insert brackets.

For example, $8x - 20 = 4(2x - 5)$
Each term is divided by 4.

To fully factorise an expression, the term outside the brackets should be the **highest common factor (HCF)** of all terms inside the brackets.

Factorise

a $6p + 15$ **b** $5x^2 + 2x$ **c** $4mn - 12m^2$

a $6p + 15$
$= 3(2p + 5)$

The HCF of $6p$ and 15 is 3.

b $5x^2 + 2x$
$= x(5x + 2)$

The HCF of $5x^2$ and $2x$ is x.

c $4mn - 12m^2$
$= 4m(n - 3m)$

The HCF of $4mn$ and $12m^2$ is $4m$.

You can always check your factorisation by expanding the brackets.

Sometimes, when you expand a pair of double brackets, some of the terms cancel out.
For example,
$$(x + 5)(x - 5) = x^2 + 5x - 5x - 25$$
$$= x^2 - 25$$

> ⬤ $(a + b)(a - b) = a^2 - b^2$
>   ▶ This is often called the **difference of two squares**.

Example

Factorise

a $x^2 - 16$ **b** $4y^2 - 25$ **c** $2p^2 - 18$

a $x^2 - 16$
$= (x + 4)(x - 4)$

You should spot that $\sqrt{16} = 4$.

b $4y^2 - 25$
$= (2y + 5)(2y - 5)$

$\sqrt{4y^2} = 2y$ and $\sqrt{25} = 5$.

c $2p^2 - 18$
$= 2(p^2 - 9)$
$= 2(p + 3)(p - 3)$

Take out the HCF of 2 and then use the difference of two squares.

Exercise 3d

1 Copy and complete these factorisations.

   a $2x + 8 = 2(x + \square)$

   b $6a - 10 = 2(3a - \square)$

   c $12 - 3p = 3(\square - p)$

   d $8b + 12 = \square (2b + \square)$

   e $5mn + 3m = m(\square + \square)$

   f $4a - 7ab = \square (4 - \square)$

   g $6x + 18xy = 6x(\square + \square)$

   h $15pq - 12p = \square (\square - 4)$

2 Fully factorise these expressions.

   a $3x + 12$ **b** $5y - 20$

   c $24a - 16$ **d** $21b - 7$

   e $14 - 8p$ **f** $10 - 10q$

   g $3x + 4xy$ **h** $12mn + 9n$

   i $4a + 12b + 20c$

   j $25a + 10ab - 15ac$

3 Fully factorise these expressions.

   a $x^2 + 2x$ **b** $3y^2 - 4y$

   c $5p^2 + p$ **d** $a^2 - ab$

   e $2n^2 + 4n$ **f** $12m^2 + 9m$

   g $5pq - 15p^2$ **h** $20x^2y - 12x$

   i $10a^2 + 20ab + 12a$

   j $6x^2 - 12x^2y + 3xy$

4 Use the difference of two squares to factorise these expressions.

   a $x^2 - 9$ **b** $x^2 - 25$

   c $y^2 - 49$ **d** $y^2 - 100$

   e $4x^2 - 25$ **f** $9a^2 - 16$

   g $4p^2 - 49$ **h** $16a^2 - 81$

   i $x^2 - 4y^2$ **j** $9m^2 - 64$

   k $(x + 1)^2 - (x - 1)^2$

   l $(x + y)^2 - (x - y)^2$

Problem solving

5 Alexander and William have each written a formula for the number of red squares, r, in a rectangle of height, n.

   Alexander wrote the formula $r = 2(n - 1)$

   William wrote the formula $r = 2n - 2$

   a   Use factorisation to explain why they are both right.

   b   **Justify** each formula by referring to the diagrams.

6 Use factorisation to simplify these algebraic fractions.

   a $\dfrac{7x + 14}{21x + 42}$ **b** $\dfrac{x^2 - y^2}{x + y}$

   c $\dfrac{a^2 - b^2}{2a - 2b}$ **d** $\dfrac{5m + 5n}{m^2 - n^2}$

7 Fully factorise these expressions.

   a $x^4 - 16$ **b** $x^4 - a^4$

Did you know?

In the 9th century, al-Khwarizmi wrote one of the first books on equations. The word algebra comes from the arabic *al-jabr* which means restoration.

3e Identities

In Maths, **identical** has a very specific meaning, more than just "looking the same".

> ● A **formula** shows the connection between two or more **variables**.

Temperatures in Fahrenheit and Celsius are connected by the formula
$F = \frac{9}{5}C + 32$ The variables are
F = temperature in °F
and C = temperature in °C.

> ● An **equation** is true for particular values of the **unknown**.

$9(x - 5) = 27$ is true only when $x = 8$ $9(8 - 5) = 27$
$4(5t - 9) = 3(t + 5)$ is true only when $t = 3$ $4(5 \times 3 - 9) = 3(3 + 5)$

> ● An **identity** is true for all values of the unknown.
> ▶ ≡ means 'is identically equal to'.

$9(x - 5) \equiv 9x - 45$ is true for all values of x
$4(5t - 9) \equiv 20t - 36$ is true for all values of t

Example

Prove that these are identities.
a $7(a - 9) - 3(a - 4) \equiv 4a - 51$ **b** $(b - 4)(b + 5) \equiv b^2 + b - 20$

a $7(a - 9) - 3(a - 4)$
$\quad = 7a - 63 - 3a + 12$
$\quad = 4a - 51$
$\quad 7(a - 9) - 3(a - 4) \equiv 4a - 51$

b $(b - 4)(b + 5)$
$\quad = b^2 - 4b + 5b - 20$
$\quad = b^2 + b + 20$
$\quad (b - 4)(b + 5) \equiv b^2 + b - 20$

> Expand the brackets to check that the left-hand side (LHS) is identical to the right-hand side (RHS).

Recognising the difference of two squares can sometimes help you with identities.

Example

Prove that these are identities.
a $(x + 5)(x - 5) \equiv x^2 - 25$ **b** $(p - q)(p + q) \equiv p^2 - q^2$

a $(x + 5)(x - 5)$
$\quad = x^2 + 5x - 5x - 25$
$\quad = x^2 - 25$
$\quad (x + 5)(x - 5) \equiv x^2 - 25$

b $(p - q)(p + q)$
$\quad = p^2 - pq + pq - q^2$
$\quad = p^2 - q^2$
$\quad (p - q)(p + q) \equiv p^2 - q^2$

Exercise 3e

1 Find the value of the required variable in each of these formulae.

a $A = l^2$

Find A when $l = 2.5$

b $P = 2l + 2w$

Find P when $l = 4\frac{1}{2}$ and $w = 2\frac{1}{4}$

c $V = \pi r^2 h$

Find V when $r = 3$ and $h = 8$

d $s = ut + \frac{1}{2}at^2$

Find s when $u = 0$, $a = 5.8$ and $t = 2$

2 Solve these equations to find the value of the unknowns.

a $3x + 5 = 23$ **b** $5(y + 4) = 35$

c $25 = 7a - 3$ **d** $20 - 4b = 8$

2 e $6(5 - p) = 18$

f $9q - 5 = 7q + 5$

g $3(m + 7) = 5m - 3$

h $22 - n = 5(n - 4)$

3 Prove that these are identities.

a $6(a + 5) \equiv 6a + 30$

b $9(b - 6) \equiv 9b - 54$

c $10k + 5(k - 7) \equiv 15k - 35$

d $3(t + 6) + 7(t + 1) \equiv 5(2t + 5)$

e $3(p - 4) + 6(p + 3) \equiv 3(3p + 2)$

f $8(q + 3) - 6(q - 3) \equiv 2(q + 21)$

g $4(2x + 3) - 3(x - 6) \equiv 5(x + 6)$

h $7(2 - y) + 8(5 + 2y) \equiv 9(y + 6)$

i $(x - 3)(x + 3) \equiv x^2 - 9$

j $(8 + y)(8 - y) \equiv 64 - y^2$

Problem solving

4 Write an algebraic expression for each area. Expand and simplify your answers using the difference of two squares.

a
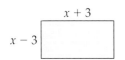

b
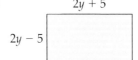

c
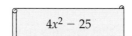

5 Match the factorised form of the difference of two squares to their expanded equivalents.

a $(x + 5)(x - 5)$

b $x^2 - 16$

c $(3x + 4)(3x - 4)$

d $9x^2 - 16$

e $(2x + 5)(2x - 5)$

f $4x^2 - 25$

g $x^2 - 25$

h $(x + 4)(x - 4)$

6 Write an identity for each of these products.

a $(a + b)^2$ **b** $(a + b)(c + d)$ **c** $(a + b)(c - d)$

7 Use the difference of two squares to do these calculations.

a $101^2 - 99^2$ **b** $1002^2 - 998^2$ **c** $10003^2 - 9997^2$

d $0.8^2 - 0.2^2$ **e** $8.7^2 - 1.3^2$ **f** $0.1^2 - 0.01^2$

 MyMaths.co.uk

3f Formulae

One of the most famous formulae is Einstein's $E = mc^2$.

● A **formula** is a relationship or rule expressed in symbols.

To find the surface area, S, of a cuboid with width w, depth d and height h you can use the formula
$$S = 2dw + 2dh + 2hw.$$
S, w, d and h are the **variables** in this formula.
You can **substitute** values into this formula to find an unknown variable.

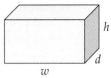

Example

Find the surface area of these cuboids.

a
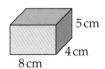
5 cm
4 cm
8 cm

b
10 mm
6 mm
8 mm

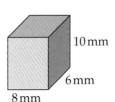

A variable is a quantity that can vary.

a $S = 2dw + 2dh + 2hw$
$= 2 \times 4 \times 8 + 2 \times 4 \times 5 + 2 \times 5 \times 8$
$= 64 + 40 + 80$
$S = 184\,cm^2$

b $S = 2dw + 2dh + 2hw$
$= 96 + 120 + 160$
$S = 376\,mm^2$

● You may need to **derive** a formula in order to solve a problem.

Derive means work out from the information given.

Example

A garden has a lawn, bordered on either side by flowerbeds.
a   Derive a formula for the area, A, of the lawn.
b   Use this formula to work out the area of the lawn if each flowerbed is 75 cm wide.

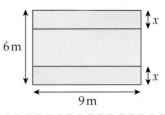

6 m

9 m

a   The width of the lawn is 6 m minus two lots of x m (one for each flowerbed).

$A = lw$
$= 9(6 - 2x)$
$A = 54 - 18x$

b $A = 54 - 18x$
$= 54 - 18 \times 0.75$
$= 40.5\,m^2$

75 cm = 0.75 m

Exercise 3f

1 If $I = \dfrac{v}{f}$ calculate I when

 a $v = 10$ and $f = 2$

 b $v = 5$ and $f = 0.5$

 c $v = 14$ and $f = -4$

2 If $s = ut + \dfrac{1}{2}at^2$ calculate s when

 a $u = 5$, $t = 2$ and $a = 10$

 b $u = -2$, $t = 3$ and $a = 4$

 c $u = 1$, $t = -6$ and $a = 2$

Take care with the order of operations!

Problem solving

3 Use the formula $S = 2dw + 2dh + 2hw$ to find the surface areas of these cuboids.

a 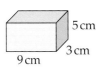 5 cm, 3 cm, 9 cm

b 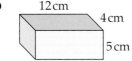 12 cm, 4 cm, 5 cm

c 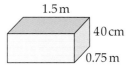 1.5 m, 40 cm, 0.75 m

4 The area of an ellipse is given by the formula

$A = \pi ab$.

Use this formula to find the area of an ellipse when

a $a = 6$ cm and $b = 4$ cm

b $a = 8.2$ cm and $b = 3.7$ cm.

Give your answers in terms of π.

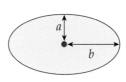

Did you know?

Hero of Alexandria was an ancient Greek mathematician and inventor. Among his many inventions are the first steam engine and the first vending machine.

5 Hero, a Greek mathematician, derived this formula for the area of a triangle with sides a, b and c

$A = \sqrt{s(s-a)(s-b)(s-c)}$

where $s = \dfrac{1}{2}(a+b+c)$.

Use Hero's formula to find the area of a triangle with sides

a 3 cm, 4 cm and 5 cm

b 4 cm, 4 cm and 6 cm.

6 A garden consists of a square lawn of length 10 m bordered by a path of width x m.

a Explain why the total length of the garden is $(10 + 2x)$ m.

b Show that the area, A, of the garden is given by the formula $A = 4x^2 + 40x + 100$.

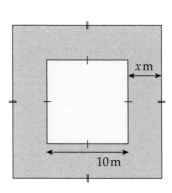

The word 'subject' has different meanings in the English language. You can be the subject of an essay, the subject of a king or the subject of an equation!

My favourite subject is history!

⬤ You **change the subject** of a formula by rearranging it.

In the formula $y = mx + c$ y is the subject.

An equivalent formula is $x = \dfrac{y - c}{m}$ x is the subject.

⬤ You can use **inverse operations** to rearrange a formula.

Example

Rearrange these formulae to make x the subject.

a $a = p + qx$

b $m = \dfrac{x + y}{n}$

Addition and subtraction are **inverses** of each other. Multiplication and division are inverses of each other.

a $a = p + qx$

This formula says 'start with x, multiply by q and add p to get a'.

$a - p = qx$ Subtract p.

$\dfrac{a - p}{q} = x$ Divide by q.

$x = \dfrac{a - p}{q}$

b $m = \dfrac{x + y}{n}$

This formula says 'start with x, add y and divide by n to get m'.

$mn = x + y$ Multiply by n.

$mn - y = x$ Subtract y.

$x = mn - y$

Example

A formula that connects temperature in °F to temperature in °C is $F = \dfrac{9}{5}C + 32$. Make C the subject of this formula and then convert 64.4 °F to °C.

$F = \dfrac{9}{5}C + 32$

$F - 32 = \dfrac{9}{5}C$ Subtract 32

$\dfrac{5}{9}(F - 32) = C$ Dividing by $\dfrac{9}{5}$ is the same as multiplying by $\dfrac{5}{9}$

$C = \dfrac{5}{9}(F - 32)$

$C = \dfrac{5}{9}(64.4 - 32) = 18$

64.4°F is 18°C

8

Exercise 3g

1 Make x the subject of these formulae.

a $x + p = q$ **b** $5y = x + y$

c $x + a = b + c$ **d** $k = x - t$

e $x - g^2 = h$ **f** $p = x - mn$

g $x + \sqrt{a} = b^2$ **h** $k + x = k + t^2$

2 Make y the subject of these formulae.

a $ay = b$ **b** $\dfrac{y}{p} = x$

c $xy = w + z$ **d** $\dfrac{y}{a} = b + c$

e $r = py + q$ **f** $ky - t^2 = x$

2   **g** $\dfrac{y}{g + h} = k$ **h** $\dfrac{y}{m} - x = n$

i $xyz = k^2$ **j** $aby = m + n$

k $p + mny = q - p$

l $y(a - b) = c^2$

3 Make a the subject of these formulae.

a $x = y(a + b)$ **b** $q(a - r) = p$

c $a(m + n) = k$ **d** $t^2(a - v) = w$

e $S = 2\pi(a + b)$ **f** $y = \dfrac{1}{2}(a - x)$

g $m = \dfrac{1}{10}(mn + a)$ **h** $k^2 = \dfrac{1}{\pi}(a - h)$

Problem solving

4 The formula $v = u + at$ connects the variables

v = final velocity a = acceleration

u = initial velocity t = time.

Find the acceleration, a, of a train which starts from rest and passes a station 22 seconds later with a velocity of 33 m/s.
(If an object starts from rest then its initial velocity is zero.)

5 The surface area of a closed cylinder is given by the formula

$S = 2\pi r(h + r)$

where r = radius and h = height.

a Show that $h = \dfrac{S - 2\pi r^2}{2\pi r}$.

b Find the height of a cylinder with a surface area of 962π cm² and a radius of 13 cm.
Leave π in your working.

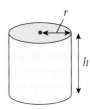

6 Match these formulae so that one is a rearrangement of the other.

a $a = b(x + y)$ **b** $x = \dfrac{a - y}{b}$ **c** $x = \dfrac{b + ay}{a}$ **d** $y = a - bx$

e $b = ax + y$ **f** $x = \dfrac{b - y}{a}$ **g** $x = \dfrac{a - by}{b}$ **h** $b = a(x - y)$

3h Changing the subject of a formula 2

If you can change the subject of these formulae, you'll be on your way to being really good at algebra!

Example

Make x the subject of these formulae.

a $b = a - x$ **b** $p = \dfrac{q}{x}$ **c** $x^2 = t$

a $b = a - x$

        $b + x = a$

          $x = a - b$

b $p = \dfrac{q}{x}$

      $px = q$

      $x = \dfrac{q}{p}$

c $x^2 = t$ Square root

     $x = \pm\sqrt{t}$ both sides.

> Remember a positive number, n, has two square roots $+\sqrt{n}$ and $-\sqrt{n}$.

Here are some guidance tips that might help:

- If the intended subject is negative, add terms to make it positive.
- If it appears in a **denominator**, multiply to undo the fraction.
- If it appears squared, you will need to take the square root.

Example

Make x the subject of these formulae.

a $m = p - qx$ **b** $\dfrac{a}{x} + b = c$ **c** $x^2 - k = t$

a $m = p - qx$

    $m + qx = p$

      $qx = p - m$

        $x = \dfrac{p - m}{q}$

b $\dfrac{a}{x} + b = c$

      $\dfrac{a}{x} = c - b$

      $a = x(c-b)$

      $x = \dfrac{a}{c - b}$

c $x^2 - k = t$

     $x^2 = k + t$

     $x = \pm\sqrt{k + t}$

Example

Make x the subject of the formula $ax + b = px + q$

    $ax - px = q - b$ Collect the terms in x on one side of the formula.

    $x(a - p) = q - b$ Factorise to isolate the term in x.

       $x = \dfrac{q - b}{a - p}$ Divide both sides by $a - p$.

> If the subject appears on both sides, collect those terms together and factorise.

   Algebra   Expressions and formulae

Exercise 3h

1 Make x the subject of these formulae.

  a $q = p - x$ **b** $g = h - x$

  c $t = mn - x$ **d** $r^2 = c - x$

  e $a = b - xy$ **f** $t = v - wx$

  g $p = k(y - x)$ **h** $r = \frac{1}{\pi}(a - x)$

2 Make y the subject of these formulae.

  a $\frac{a}{y} = c$ **b** $t = \frac{k}{y}$

  c $\frac{p}{y} = \frac{q}{r}$ **d** $e = \frac{x}{yz}$

  e $\frac{t}{y} + r = s$ **f** $a = \frac{b}{y} - c$

  g $k^2 = t + \frac{x}{y}$ **h** $\frac{c}{y} - n = -m$

  i $\frac{f}{g + y} = h$ **j** $x = \frac{a}{y - b}$

  k $p = q - \frac{r}{y}$ **l** $n = x - \frac{m}{y}$

3 Make p the subject of these formulae.

  a $p^2 = t$

  b $p^2 - q = k$

  c $m = p^2 - n^2$

  d $\sqrt{p} = x$

  e $ap^2 = b$

  f $p^2x = y$

  g $t - p^2 = k$

  h $f = g - p^2$

4 Rearrange these formula with x on both sides to make x the subject.

  a $ax + b = cx + d$

  b $px + q = r - tx$

  c $hx - g = m - nx$

  d $p - qx = r - sx$

Problem solving

5 The surface area of an isosceles triangular prism is given by this formula.

$A = bh + 2ls + lb.$

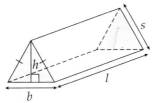

  a Find the surface area, A, of an isosceles triangular prism with $b = 6\,cm$, $h = 4\,cm$, $l = 10\,cm$ and $s = 5\,cm$.

  b Rearrange the formula $A = bh + 2ls + lb$ to make

    i s the subject   **ii** b the subject.

6 The diagram shows a cylinder with radius r and height h.

  a Try to find a formula for the **volume** of the cylinder.

  b Rearrange your formula to make r the subject.

7 The formula for the volume of a sphere is

$V = \frac{4}{3}\pi r^3$

Rearrange the formula to make r the subject.

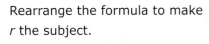

3 MySummary

Check out

You should now be able to ...

Test it ➡

Questions

✓ Know and use the index laws.	7	1 – 4
✓ Multiply brackets in two linear expressions.	8	5
✓ Factorise expressions by taking common factors.	8	6 – 8
✓ Derive simple identities, including expansion of two linear brackets.	8	9
✓ Derive formulae and substitute values in formulae to find unknown variables.	8	10
✓ Change the subject of a formula.	8	11, 12

Language	Meaning	Example
Index/indices power	A number written as a power has a base and an index (plural indices) or power.	6^3 The base is 6 and the index is 3 $6^3 = 6 \times 6 \times 6$
Difference of two squares	An expression of the form $x^2 - b^2$ that factorises to $(x + a)(x - b)$.	$x^2 - 25 = (x + 5)(x - 5)$
Variable	An unknown quantity that can take different values.	In the formula for the area of a circle $A = \pi r^2$ the variable is r
Formula	Shows the connection between several variables.	Area of a circle, $A = \pi r^2$
Equation	An algebraic statement which has particular solutions.	$4x - 7 = 5$ has only one solution $x = 3$
Identity	An algebraic statement which is true for all solutions.	$2(x + 4) \equiv 2x + 8$ for every possible value of x
Change the subject	Rearrange an equation so that a different variable is "on its own".	$v = \dfrac{b^2}{k}$ rearranged to make b the subject gives $b = \sqrt{vk}$

Algebra Expressions and formulae

1 Evaluate without using a calculator
   a 10^3 **b** $(-2)^3$

2 Simplify these expressions.
   a $5a^3 \times 4a^5$ **b** $21b^7 \div 7b^2$
   c $(3c^6)^3$ **d** $\dfrac{5d^8 \times 6d}{15d^3}$

3 Evaluate without using a calculator
   a 25^0 **b** $16^{\frac{1}{2}}$
   c 12^{-1} **d** $100^{-\frac{1}{2}}$

4 Simplify these expressions.
   a $e^{\frac{1}{2}} \times e^{\frac{1}{2}}$ **b** $(f^{\frac{1}{2}})^3$
   c $g^3 \div g^{-2}$ **d** $\dfrac{h^2 \times h^{-\frac{1}{4}}}{h^{\frac{3}{4}}}$

5 Expand the brackets and simplify your answers.
   a $(a + 2)(a + 1)$
   b $(b + 7)(b - 4)$
   c $(c - 8)(c - 3)$
   d $(d - 5)^2$
   e $(2e + 3)(e - 4)$
   f $(f + 5)(f - 4) - (f - 4)^2$

6 Factorise these expressions.
   a $4g - 20$
   b $32 - 24h$
   c $15mn + 20m$
   d $12xy + 4y$

7 Factorise these expressions.
   a $x^2 + 7x$ **b** $5x^2 + 10xy$
   c $4y^3 + 6y^2 - 2y$ **d** $12y^3 - 15y^2z$

8 Factorise these expressions.
   a $p^2 - 9$ **b** $4q^2 - 1$
   c $r^2 - 81s^2$ **d** $100t^2 - u^2$

9 Are each of these identities correct?
   a $3(5a + 2) \equiv 15a + 6$
   b $(b - 4)^2 \equiv b^2 + 16$
   c $3(4c - 6) + 2(7 - c) \equiv 2(5c - 2)$

10 Give a formula for the shaded area in terms of π and a.

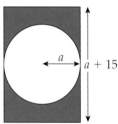

11 Make x the subject of these formulae.
   a $x + 2a = b$ **b** $ax - b = c$
   c $\dfrac{x}{d + e} = 3f$ **d** $a + bx = c - a$

12 Make y the subject of these formulae.
   a $\dfrac{a}{y} = 2b$ **b** $c = \dfrac{3}{y + d}$
   c $\dfrac{2}{y} + e = f$ **d** $a - by^2 = 3c$

What next?

Score			
	0 – 4		Your knowledge of this topic is still developing. To improve look at Formative test: 3C-3; MyMaths: 1033, 1178, 1045, 1150, 1247, 1155, 1157, 1186, 1187, 1171 and 1170
	5 – 10		You are gaining a secure knowledge of this topic. To improve look at InvisiPen: 185, 221, 222, 251, 252, 254, 255 and 256
	11 – 12		You have mastered this topic. Well done, you are ready to progress!

3a

1   Simplify these expressions leaving your answers in index form.

a $2^5 \times 2^3$
b $4^2 \times 4^7$
c $3^6 \div 3^4$
d $\dfrac{8^{10}}{8^7}$

e $(5^3)^4$
f $x^5 \times x^4$
g $\dfrac{a^6}{a^5}$
h $(n^5)^2$

2   Simplify these expressions leaving your answers in index form.

a $4x^3 \times 3x^2$
b $\dfrac{6y^7}{2y^5}$
c $(2p^6)^3$
d $\dfrac{2n^5 \times 6n}{4n^3}$

3b

3   Evaluate these powers without a calculator.

a 9^0
b 3^{-2}
c 5^{-3}
d 12^0

e $36^{\frac{1}{2}}$
f $100^{\frac{1}{2}}$
g $27^{\frac{1}{3}}$
h $64^{\frac{1}{3}}$

4   Simplify these expressions leaving your answers in index form.

a $3^4 \times 3^{-7}$
b $10^{-2} \div 10^3$
c $(9^5)^{-4}$
d $\dfrac{7^3}{7^9}$

3c

5   Expand these double brackets and simplify.

a $(x + 2)(x + 3)$
b $(y + 5)(y + 2)$
c $(t + 7)(t + 6)$
d $(a + 4)^2$
e $(b + 5)(b - 3)$
f $(d + 7)(d - 5)$
g $(m + 2)(m - 4)$
h $(n + 10)(n - 10)$
i $(p - 3)(p - 4)$
j $(q - 5)(q - 1)$
k $(f - 3)(f - 9)$
l $(k - 3)^2$

6   Write an algebraic expression for the area of each circle.
Expand the brackets and simplify your answer.

a

b

3d

7   Fully factorise these expressions.

a $4x + 8$
b $12y - 9$
c $15p - 10$
d $12 - 12q$

e $7m + 11mn$
f $20ab - 24b$
g $12x + 15y + 9z$
h $18xy - 27yz + 9y$

8   Fully factorise these expressions.

a $x^2 + 10x$
b $y^2 - 5y$
c $mn + n^2$
d $a^2b + a$

e $2k^2 - 5k$
f $7t^2 + 14t$
g $16xy^2 - 20y$
h $9p^2q + 15pq - 12p$

9 Solve these equations to find the value of the unknown.

a $5x + 7 = 22$ **b** $26 = 3y - 4$ **c** $4(a + 6) = 28$

d $3(2b - 1) = 21$ **e** $18 - 2p = 4$ **f** $5(7 - 3q) = 20$

g $10m - 11 = 7m - 2$ **h** $17 - 2n = 3(n - 1)$ **i** $16 + 3c = 5(c + 4)$

10 Copy and complete these identities.

a $5(x - 9) \equiv \square - 45$ **b** $8(y + 4) \equiv \square + \square$

c $6a + 3(a - 4) \equiv \square - \square$ **d** $9(b + 2) - \square \equiv 4b + \square$

e $4(m + 3) + 6(m - 1) \equiv 2(\square + \square)$ **f** $8(n - 1) - 5(n - 4) \equiv \square(\square + 4)$

11 The volume of a cone is given by the formula
$V = \frac{1}{3}\pi r^2 h$
Use the formula to find the volume
of a cone when

a $r = 2$ and $h = 6$

b $r = 5$ and $h = 12$.

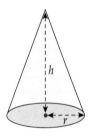

Leave your answer
in terms of π.

12 In mechanics a formula connecting the
initial velocity, u, the final velocity, v, the acceleration, a, and
the distance travelled is
$$v^2 = u^2 + 2as$$
Find   **a** v if $u = 1$, $a = 7$ and $s = 10$

b a if $v = 10$, $u = 5$ and $s = 15$

c u if $v = 13$, $a = 3$ and $s = 4$

13 Make x the subject of these formulae.

a $x + a = b$ **b** $x + y + z = k$ **c** $q = x - p$

d $x + t^3 = r$ **e** $nx = m$ **f** $\dfrac{x}{ab} = c$

g $xy = a + b + c$ **h** $ax - b = k$ **i** $p(x + y) = r$

j $k = \pi(x - t)$ **k** $px = a + qx$

14 Make k the subject of these formulae.

a $a - k = b$ **b** $t = xy - k$ **c** $p - kt = q$

d $m = n(t - k)$ **e** $\dfrac{a}{k} = b$ **f** $x = \dfrac{y}{kw}$

g $y = \dfrac{a}{k} + b$ **h** $\dfrac{d}{k + f} = e$ **i** $k^2 = x$

j $ak^2 = b$ **k** $(pk)^2 = q$ **l** $a\sqrt{k} + b = c$

Bikes are ingeniously simple vehicles which are very efficient at getting us around quickly and cheaply. This case study shows how bikes have developed over the years into the sophisticated machines they are today.

Task 1

The pedals of a penny-farthing bicycle were fixed directly to the front wheel so the wheel turned once for every turn of the pedals. The larger the wheel, the further the bike travelled for each turn.

This penny-farthing has a wheel diameter of 1.5 m.

a How far would the bike travel for one turn of the pedals?
  Remember: $C = \pi d$

b How many turns of the pedals would be needed to travel 1 km?

Task 2

If you remember riding a tricycle like this, you will know that you had to pedal quite quickly even at low speeds!

a With a 30 cm diameter front wheel, how many turns of the pedals would be needed to travel 500 m?

b Why does a child have to pedal quickly on this type of tricycle?

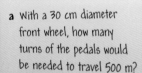

Task 3

As bikes developed, their wheels became smaller and a crank and chain drive was used. The larger front sprocket means that the wheel turns several times for each turn of the pedals.

Imagine that there are 40 teeth on the front sprocket and 20 teeth on the rear sprocket. Then each turn of the front sprocket turns the rear sprocket twice. So each turn of the pedals turns the wheels twice.

How many times would the wheels turn for these sprocket combinations?

a 36 teeth front, 12 teeth rear
b 42 teeth front, 28 teeth rear

Task 4

Most bikes now have several gears which select different numbers of teeth on the front and rear sprockets.
This allows you to alter the number of turns of the wheels for each turn of the pedals.

a A 7-speed touring bike has wheel diameter 700 mm.
   Find the distance travelled for each turn of the wheel, giving your answer to 2 d.p.

b Copy and complete this table, which shows different gear selections for the bike.
   Give your answers to 2 d.p. where appropriate.

number of teeth front sprocket	number of teeth rear sprocket	number of turns of the wheel per turn of the pedals	distance travelled per turn of the pedals (m)
48	12	4	
48	14		
48	16		6.60
48	18		
48	20		
48	24		
48	28		

c With these gear selections, what is the fewest number of turns of the pedal that would be needed to travel 1 km?

Task 5

When riding comfortably, a cyclist makes between 40 and 90 turns of the pedals per minute.
Look again at the table in task 4.
What range of speeds will a cyclist travel at in each gear? Give your answers in kilometres per hour.

a Describe the amount of overlap between the speed ranges in different gears.

b Why would this be a good thing?

c What would the speed ranges be in miles per hour?
   1 km $\approx \frac{5}{8}$ mile

d What would the speed ranges be if the cyclist were using their 32 teeth front sprocket?

Extension

If you have a bike, estimate how quickly you turn the pedals and work out the speed ranges for your gears.

4 Fractions, decimals and percentages

Introduction

Fractals are at the cutting edge of modern mathematics, but are based on the simple idea of fractions. A fractal is a picture that, as you zoom in to any part of it, looks like the original picture! If the picture is infinitely detailed, then you should be able to zoom in forever and you will end up with what you started with.

What's the point?

Fractals are vitally important in biology and medicine for understanding the organs of the human body at microscopic level; they are also useful for describing complex and irregular real world objects such as clouds, coastlines and trees. They are frequently used in computer games to generate 'real-life' environments.

Objectives

By the end of this chapter, you will have learned how to ...

- ◉ Add, subtract, multiply and divide fractions.
- ◉ Convert decimals to fractions and fractions to decimals.
- ◉ Find percentage increases and decreases.
- ◉ Solve percentage problems using a decimal multiplier.
- ◉ Calculate a repeated percentage increase and decrease.

Check in

1 Evaluate, giving your answer as a fraction in its simplest form.

   a $\dfrac{2}{7} + \dfrac{1}{3}$ **b** $\dfrac{4}{7} - \dfrac{2}{5}$ **c** $\dfrac{2}{3} \times \dfrac{4}{5}$ **d** $\dfrac{2}{3} \div \dfrac{4}{5}$

2 Work out these.

   a $\dfrac{3}{7}$ of £84 **b** $\dfrac{3}{5}$ of 28 kg

   c Increase £60 by 8% **d** Decrease £350 by 3.5%

3 Write each of these ratios in their simplest form.

   a $18:24$ **b** $169:65$ **c** $350\,g:1.5\,kg$ **d** $300\,mm:0.45\,m$

Starter problem

Look at this sequence.

  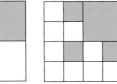

Stage 0 Stage 1 Stage 2

Describe a rule for making the next pattern in the sequence.

Draw the next pattern in the sequence.

What additional fraction of the new square is now shaded?

What is the total fraction of your new square that is now shaded?

Investigate this sequence.

● You can add or subtract fractions with different denominators by first writing them as **equivalent fractions** with the same denominator.

Calculate $3\frac{3}{4} + 1\frac{5}{6}$

Rewrite both fractions as improper fractions.

$3\frac{3}{4} = \frac{15}{4}$ $1\frac{5}{6} = \frac{11}{6}$

$\overset{\times 3}{\frac{15}{4} = \frac{45}{12}}$ $\overset{\times 2}{\frac{11}{6} = \frac{22}{12}}$ Rewrite both fractions as equivalent fractions with the same denominator.
The lowest common denominator (LCM) of 4 and 6 is 12.

$\frac{45}{12} + \frac{22}{12} = \frac{45 + 22}{12}$ Only add the numerators.

$= \frac{67}{12} = 5\frac{7}{12}$

▲ How often do you use fractions?

● You can multiply fractions by integers or by other fractions.

Calculate $2\frac{2}{5} \times 1\frac{1}{9}$

$2\frac{2}{5} = \frac{12}{5}$ $1\frac{1}{9} = \frac{10}{9}$ Rewrite both fractions as improper fractions.

$\frac{12}{5} \times \frac{10}{9} = \frac{\overset{4}{\cancel{12}}}{\underset{1}{\cancel{5}}} \times \frac{\overset{2}{\cancel{10}}}{\underset{3}{\cancel{9}}} = \frac{4 \times 2}{1 \times 3}$ Cancel any common factors.

$= \frac{8}{3} = 2\frac{2}{3}$

When you multiply a pair of fractions, both the numerators and denominators are multiplied together.

● You can divide fractions by integers or by other fractions using the **reciprocal**.

Calculate $1\frac{3}{5} \div \frac{4}{7}$

$\frac{4}{7} \rightarrow \frac{7}{4}$ Change the divisor to its reciprocal.

$1\frac{3}{5} = \frac{8}{5}$

$\frac{8}{5} \div \frac{4}{7} = \frac{8}{5} \times \frac{7}{4}$ Change the division into a multiplication.

$= \frac{56}{20} = \frac{14}{5} = 2\frac{4}{5}$

To divide by a fraction you multiply by its reciprocal. You find the reciprocal of a fraction by turning it upside down.

7

Exercise 4a

1 Work out the following, leaving your answers as fractions in their simplest form.

   a $\frac{5}{8} + \frac{3}{4}$ **b** $\frac{2}{3} - \frac{3}{8}$ **c** $1\frac{3}{5} + 1\frac{1}{6}$ **d** $2\frac{7}{10} - 1\frac{6}{7}$

   e $2\frac{5}{7} + 3\frac{7}{8}$ **f** $1\frac{4}{9} + 2\frac{5}{6}$ **g** $2\frac{1}{5} - 1\frac{5}{7}$ **h** $2\frac{7}{12} + 3\frac{5}{8}$

2 Calculate each of the following, using cancellation where appropriate.

   a $8 \times \frac{3}{4}$ **b** $12 \times 1\frac{7}{10}$ **c** $24 \times 2\frac{5}{18}$ **d** $\frac{5}{9} \times \frac{6}{11}$

   e $1\frac{3}{8} \times \frac{6}{11}$ **f** $2\frac{4}{7} \times 1\frac{13}{15}$ **g** $1\frac{3}{25} \times 1\frac{4}{21}$ **h** $2\frac{13}{16} \times 2\frac{2}{27}$

3 Calculate each of the following, leaving your answer in its simplest form.

   a $3 \div \frac{3}{7}$ **b** $12 \div \frac{6}{11}$ **c** $8 \div 1\frac{3}{5}$ **d** $1\frac{3}{8} \div \frac{5}{6}$

   e $2\frac{4}{9} \div \frac{2}{3}$ **f** $2\frac{2}{5} \div 1\frac{1}{8}$ **g** $3\frac{5}{9} \div 1\frac{3}{5}$ **h** $2\frac{3}{16} \div 2\frac{11}{12}$

Problem solving

4 a A picture frame in the shape of a rectangle has a length of $8\frac{7}{16}$ inches and a width of $5\frac{3}{4}$ inches.
Calculate the perimeter of the picture frame.

   b Declan takes a train journey from Kwale to Horsel.
The total distance is 30 km.
How far is it from Gornt to Horsel?

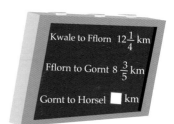

Kwale to Fflorn $12\frac{1}{4}$ km

Fflorn to Gornt $8\frac{3}{5}$ km

Gornt to Horsel ☐ km

5 a A book cover is $6\frac{3}{5}$ inches wide and $12\frac{4}{7}$ inches long.
What is the area of the book cover?

   b A bag of peanuts weighs $\frac{3}{5}$ of a kg. During October Kirsty eats $3\frac{1}{2}$ bags of peanuts. How many kilograms is that?

6 a How many sixths are there in $3\frac{1}{3}$?

   b How many quarters are there in $2\frac{1}{5}$?

7 a Here is a sequence of fractions: $\frac{1}{2}, \frac{2}{3}, \frac{3}{4}, \frac{4}{5}, \frac{5}{6}, \ldots$
Multiply the fractions in the sequence.
$\frac{1}{2} \times \frac{2}{3}, \frac{1}{2} \times \frac{2}{3} \times \frac{3}{4}, \ldots$
Write down what you notice.

   b Investigate what happens when you multiply or add other fraction sequences. For example,
$\frac{1}{2} \times \frac{3}{4} \times \frac{5}{6} \times \frac{7}{8} \ldots$ or
$\frac{1}{2} + \frac{1}{8} + \frac{1}{32} + \frac{1}{128} \ldots$ or
$\frac{1}{2} \times \frac{3}{4} \times \frac{5}{8} \times \frac{7}{16} \ldots$

Did you know?

The Greek philosopher Zeno of Elea was famous for his paradoxes. Consider a door closing. First it closes half way and then half of the remaining distance and then half the distance again, repeating forever. Does it ever close?

It is often easier to use a fraction rather than a decimal.

$\frac{1}{3} = 0.3333...$ $\frac{1}{7} = 0.142857142...$

⬤ You can convert a **terminating decimal** into a fraction using place value.

Example

Convert these decimals into fractions in their simplest form.

a   0.4   **b**   0.325

- -

a $0.4 = \frac{4}{10}$ **b** $0.325 = \frac{325}{1000}$

$\quad\quad = \frac{2}{5}$ $\quad\quad\quad = \frac{13}{40}$

> Divide top and bottom by their HCF to simplify.

⬤ You can convert a **recurring decimal** to a fraction using algebra.

Example

Convert these decimals into fractions in their simplest form.

a $0.\dot{3} = 0\,33333...$ **b** $0.\dot{2}\dot{7} = 0.272727...$

- -

a   Let $x = 0.33333...$ **b**   Let $x = 0.272727...$

$10x = 3.33333...$ 1 digit recurs, multiply $100x = 27.272727...$ 2 digits recur, multiply
$\quad\quad\quad\quad\quad\quad$ by 10. $\quad\quad\quad\quad\quad\quad\quad$ by 100.

$9x = 3$ Subtract x from $10x$. $99x = 27$ Subtract x from $100x$.

$x = \frac{3}{9} = \frac{1}{3}$ $x = \frac{27}{99} = \frac{3}{11}$

⬤ The **reciprocal** of a number is the result of dividing it into 1.

Example

Find the reciprocal of these numbers.

a   6   **b**   0.4   **c** $\frac{1}{6}$ **d** $\frac{2}{5}$

- -

First write the number as a fraction.

a $6 = \frac{6}{1}$ **b** $0.4 = \frac{4}{10}$ **c** $\frac{1}{6}$ **d** $\frac{2}{5}$

$\quad\quad\quad\quad\quad\quad\quad = \frac{2}{5}$

To find the reciprocal of a fraction turn it upside down.

$\quad\quad\frac{1}{6}\quad\quad\quad\quad\frac{5}{2}\quad\quad\quad\quad\frac{6}{1} = 6\quad\quad\quad\quad\frac{5}{2}$

> You can use the x^{-1} key on your calculator to calculate the reciprocal.

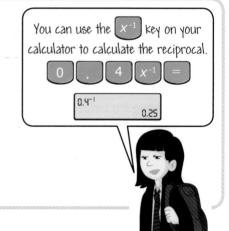

8

Exercise 4b

1 Write each decimal as a fraction in its simplest form.

a	0.3	**b**	1.24	**c**	0.56
d	0.78	**e**	0.355	**f**	0.045
g	1.625	**h**	0.435	**i**	2.045
j	1.555	**k**	0.055	**l**	0.008

2 Write these fractions as decimals without using a calculator.

a $\frac{7}{10}$ **b** $\frac{13}{20}$ **c** $\frac{16}{25}$

d $\frac{43}{50}$ **e** $\frac{35}{25}$ **f** $\frac{9}{5}$

g $\frac{27}{20}$ **h** $\frac{53}{40}$ **i** $\frac{19}{16}$

j $\frac{165}{80}$ **k** $\frac{3}{800}$ **l** $\frac{33}{800}$

3 Change these fractions into decimals using division. Use an appropriate method. Give your answers to 5 sf.

a $\frac{1}{3}$ **b** $\frac{3}{7}$ **c** $\frac{7}{11}$

d $\frac{6}{13}$ **e** $\frac{4}{17}$ **f** $\frac{2}{7}$

4 Write these recurring decimals as fractions in their simplest form.

a	0.666...	**b**	0.111...
c	0.2323...	**d**	0.4545...
e	0.354354...	**f**	0.801801...
g	0.729729...	**h**	0.162162...

5 Find the reciprocals of these numbers without using a calculator.
Leave your answers in the most appropriate form.

a 7 **b** $\frac{1}{4}$

c 0.2 **d** $\frac{2}{3}$

e 2.6 **f** $\frac{1}{13}$

g $\frac{72}{83}$ **h** 26.2

6 Write these recurring decimals as fractions in their simplest form.

a	0.1666...	**b**	0.53333...
c	0.58333...	**d**	0.291666...

Problem solving

7 Jose thinks of a number between 0 and 100. He finds the reciprocal of the number and rounds his answer to 4 significant figures. He writes down his answer as 0.013*1.
Find Jose's number and the missing digit.

8 Kelvin wants to draw the graph of $y = \frac{1}{x}$.
This is called the **reciprocal function**. He completes a table of values.

x	0.5	1	1.5	2	2.5	3
y						

a Copy and complete the table.
b Draw the graph of $y = \frac{1}{x}$.
c Use some more values for x which are smaller than 1. Plot these points on your graph. Describe what is happening to your graph as x gets smaller.
d Use some larger values for x which are greater than 3. Plot these points on your graph. Describe what is happening to your graph as x gets larger.
e Try some negative values for x. Describe what happens.

Did you know?

The graph of $y = \frac{1}{x}$ is an example of a **conic section** called a **hyperbola**. You make a hyperbola by slicing through a cone at a shallow angle to its axis.

MyMaths.co.uk

🔍 1063, 1066 **SEARCH**

Percentages are used all the time, for example to add VAT to bills.
You can calculate a percentage of an amount using mental, written
and calculator methods.

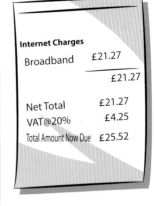

Internet Charges

Broadband	£21.27
	£21.27
Net Total	£21.27
VAT@20%	£4.25
Total Amount Now Due	£25.52

Example

Calculate 5% of £38.

Using a mental method.
10% of £38 = £3.80
so 5% of £38 = £1.90

Using an equivalent fraction.

5% of £38 = $\frac{5 \times 38}{100}$ = $\frac{190}{100}$ = £1.90   Change the % into a fraction and multiply.

Using the decimal equivalent.

5% of £38 = 5 ÷ 100 × 38   Change the % into a decimal
          = 0.05 × 38 = £1.90   and multiply.

● You can calculate a **percentage increase** or **decrease** in a
  single calculation using an equivalent decimal.

Example

a Increase £70 by 12%   **b** Decrease 250g by 12%

a New price
= (100 + 12)% of £70
= 112% of £70
= 1.12 × £70
= £78.40

b New weight
= (100 − 12)% of 250g
= 88% of 250g
= 0.88 × 250
= 220g

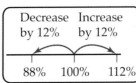

Decrease   Increase
by 12%   by 12%

88%   100%   112%

● You can express a **proportional** change as a percentage of the
  original amount.

Percentage change = $\frac{\text{Change}}{\text{Original amount}}$ × 100%

Example

A chocolate bar is decreased in weight from 250g to 220g.
What is the percentage decrease in weight?

Decrease in weight   = 250g − 220g
                      = 30g

Percentage decrease = $\frac{30}{250}$ × 100%
                    = 12%

Exercise 4c

1 Calculate these percentages using a suitable method.

   a 15% of £35   **b** 95% of 180 kg   **c** 11% of 85 km.   **d** 6% of £37

   e 99% of 2.5 m   **f** 17.5% of £2450   **g** 3.5% of £28 000   **h** 0.5% of 55 kg

2 a From every 800 tins of beans tins made in a factory, 11 are rejected.

     What proportion of tins are rejected?

   b Nadeem scores 53 out of 60 in his Maths exam.

     What proportion of the test did he answer incorrectly?

3 Use an appropriate method to work out the following.

   a Increase £75 by 15%   **b** Decrease £142 by 15%

   c Increase 12.5 km by 11%   **d** Increase 1.2 million by 4.5%

   e Decrease 30 cm by 78%   **f** Decrease 2900 kJ by 13.5%

Problem solving

4 Match each of these statements with the correct mathematical calculation.

A Increase £75 by 30% **B** 30% of £75	
C Decrease £75 by 30% **D** Increase £75 by 3%	
E 3% of £75 **F** Decrease £75 by 3%	
G Increase £75 by 0.3% **H** 0.3% of £75	

a 0.03×75	**b** 1.03×75
c 0.3×75	**d** 1.3×75
e 1.003×75	**f** 0.97×75
g $0.003 \times £75$	**h** 0.7×75

5 a A DVD costs £19. In a sale the price is reduced to £12.

     What is the percentage reduction?

   b A skirt normally costs £25. In a sale the price is reduced by 35%.

     What is the sale price of the skirt?

   c Hector's weekly wage is increased from £320 to £329.60.

     What is the percentage increase in Hector's wage?

6 These are the prices of four items in 1982 and in 2007.

   Which item has increased in price by the greatest proportion?

   Explain and justify your answer.

	1982	2007
Cinema ticket	£1.65	£6.50
Pint of milk	20p	36p
Weekly food bill	£32.20	£84.82
Chocolate bar	16p	41p

Did you know?

£1 in 1982 is equivalent to £2.99 in 2012. Inflation is the rate of increase in prices. The rate of inflation from 1982 to 2012 is 199%.

You can use percentage changes to solve problems.

> ● You can use a decimal multiplier to solve percentage problems in reverse.

In a sale a tennis racket is reduced in price by 35%.
The racket now costs £18.85.
What was the original price of the racket?

Sale price of racket = (100 − 35)% of the original price
= 65% of original price

To work out 65% of the original price you multiply by 0.65

$$\times 0.65 \longrightarrow$$

Original price   decrease by 35%   Sale price

$$\longleftarrow \div 0.65$$

= □   ÷ 0.65   £18.85

Decimal multiplier = 100% − 35% = 65%
= 0.65

Original price of racket = Sale price ÷ 0.65
= £18.85 ÷ 0.65
= £29

John has taken up body building.
He gains 12% of his body weight and now weighs 89.6 kg.
What was his original weight?

His weight has increased by 12%

$$\times 1.12 \longrightarrow$$

Original weight   Increase by 12%   New weight

= □   ÷ 1.12   89.6 kg

Decimal multiplier = 100% + 12% = 112%
= 1.12

Original weight = new weight ÷ 1.12
= 89.6 ÷ 1.12
= 80 kg

You can check using percentage change

$$= \frac{\text{Change in weight}}{\text{Original weight}}$$

$$= \frac{89.6 - 80}{80}$$

$$= \frac{9.6}{80} = 0.12 = 12\% \checkmark$$

Exercise 4d

1 These are final prices. What was the original price in each case?

   a £546 after a reduction of 35%   **b** £51.87 after an increase of 14%

   c £33.72 after a reduction of 76%   **d** £13 645.06 after an increase of 43%

   e £145 after an increase of 150%   **f** £8946 after an increase of 5%

Problem solving

2 Match each of these questions with the correct mathematical calculation.

   a A pair of trainers costs £30. The price is increased by 20%. Calculate the new price.

   b In a sale all prices are reduced by 20%. A pair of trainers costs £30 in a sale. Calculate the old price of the trainers.

   c A pair of shoes costs £30. The price is reduced by 20%. Calculate the new price.

   d In a shop all prices are increased by 20%. A pair of slippers now costs £30. What was the old price of the slippers?

   A 30 ÷ 0.8

   B 30 ÷ 1.2

   C 30 × 0.8

   D 30 × 1.2

3 a Felippe bought a computer in a sale and saved £49. The label said that it was a 20% reduction. What was the original price of the computer?

   b A packet of sweets is increased in size by 23%. The new packet weighs 184.5 g. What was the weight of the original packet?

4 A company advertises its new packet of sweets claiming that it is 27% bigger. The new packet contains 19 sweets. How many sweets were there in the original packet? Comment on your answer.

5 Mr P. Centage owns a shop. He always puts up his prices by 5% on the 1st April each year. In 2009 the price of a chocolate bar in his shop was 80p.

   a How much will the chocolate bar cost in two years' time?

   b What did the chocolate bar cost last year?

   c What did the chocolate bar cost two years ago?

   d Investigate the price of the chocolate bar in different years.

When you borrow money it is important to know how much interest you will pay.

⬤ You can calculate a **repeated percentage** increase using a **decimal multiplier**.

Example

Victor invests £3000 in a bank with an annual **interest** rate of 4%. How much money will he have after 3 years?

Each year the money grows by 4%.

$$£3000 \xrightarrow{\times 1.04} £3120 \xrightarrow{\times 1.04} £3244.80 \xrightarrow{\times 1.04} £3374.592$$

Increase by 4%   Increase by 4%   Increase by 4%

After three years he will have £3374.59 (2 dp)

You can also perform the same calculation in one step.

£3000 × $(1.04)^3$ = £3374.59   1.04 × 1.04 × 1.04 = 1.04^3

This is called compound interest

At the end of each year the bank works out 4% of the money in the account, called the interest, and adds this on.

⬤ You can calculate a repeated percentage decrease using a decimal multiplier.

Example

Violet buys a new van for £20000.
Each year the value of the van depreciates by 12%.
What will be the value of the van in two years' time?

Each year the value of the van decreases by 12%.

$$£20000 \xrightarrow{\times 0.88} £17600 \xrightarrow{\times 0.88} £15488$$

Decrease by 12%   Decrease by 12%

After two years the van is worth £15488.

You can also perform the same calculation in one step.

£20000 × $(0.88)^2$ = £15488   0.88 × 0.88 = 0.88^2

Most cars go down in value each year. This is called **depreciation**.

Exercise 4e

1 Write the decimal multiplier for each of these changes.

   a Value increasing by 4% for one year.

   b Value increasing by 4% for four years.

   c Value increasing by 5% for five years.

   d Value depreciating by 8% for six years.

   e Value depreciating by 1% for 20 years.

   f Value depreciating by 50% for 20 years.

   g Value increasing by 1% for a century.

Problem solving

2 **a** Ruby puts £8000 into a bank account. Each year the bank pays interest at 7%. Work out the amount of money in Ruby's bank account after two years.

   b Shahid buys a new car for £12 000. Each year his car depreciates in value by 15%. What will be the value of Shahid's car in three years' time?

3 Work out the value of each of these items after the number of years stated.

Item	Cost (£)	Time (years)	Percentage change each year
Car	24 500	3	Decrease 7%
House	125 000	2	Increase 8%
TV	250	5	Decrease 12.5%
Savings	15 000	4	Increase 3.29%

4 **a** Teresa invests £85 000 in a bank account. The bank pays 6% interest a year. How many years would it take for Teresa's money to be worth more than a million pounds?

   b Bertie owes £1800 on his credit card. Each month the credit card company adds 2.6% interest on his outstanding balance to his bill. Bertie manages to pay £80 a month to the credit card company. How long will it take him to pay off the whole bill?

Did you know?

If you invest your savings at 5% compound interest your money will double in just over 14 years.

5 Bea borrows £90 000 from a bank to buy a house. The bank charges 6% interest a year. Bea wants to know how much money she will have to pay back after 25 years. She uses a spreadsheet to help her investigate the problem.

   a Set up a spreadsheet to investigate the problem.

   b How much money will Bea owe in 25 years?

   c Plot a graph of time in years against the money owed. Write what you notice.

   d Investigate changing the interest rate.

   e Investigate what happens if Bea starts to pay back some money at the end of each year. Find how much she has to pay back each year so that at the end of the 25 years she owes the bank no money?

	A	B	C	D
	Year	Amount	Percentage change	New amount
1	Year	Amount	Percentage change	New amount
2	1	90 000	1.06	95 400
3	2	95 400	1.06	101 124
4	3	101 124	1.06	107 191
5	4	107 191	1.06	113 623
6	5	113 623	1.06	120 440
7	6	120 440		

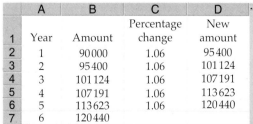

Check out

You should now be able to ...

Test it ➡
Questions

✓ Add, subtract, multiply and divide fractions.	7	1, 2
✓ Convert decimals to fractions and fractions to decimals.	8	3, 4
✓ Find percentage increases and decreases.	8	5 – 7
✓ Solve percentage problems using a decimal multiplier.	8	8, 9
✓ Calculate a repeated percentage increase and decrease.	8	10, 11

Language	Meaning	Example
Equivalent fractions	Equal fractions with different denominators.	$\frac{1}{2} = \frac{3}{6} = \frac{25}{50}$
Terminating decimal	A decimal with a finite number of digits.	0.25
Recurring decimal	A decimal with an infinite number of digits which repeat in a pattern.	0.12121212... 0.33333...
Percentage increase/decrease	Changes expressed as fractions of 100.	$35\% = \frac{35}{100} = 0.35$
Proportional change	An increase or decrease written as a fraction or % of the original amount.	An increase from 80 to 100 is a proportional increase of $\frac{20}{80} = 25\%$
Decimal multiplier	The decimal equivalent to a % change.	To increase a value by 12% you multiply by 1.12

1   Calculate these, writing your answers as improper fractions.

   a $3\frac{3}{5} + 1\frac{2}{7}$ **b** $4\frac{7}{9} - 2\frac{1}{4}$

   c $1\frac{7}{12} + 2\frac{4}{15}$ **d** $3\frac{7}{10} - 1\frac{11}{12}$

2   Calculate these, giving your answers in their simplest form.

   a $8 \times \frac{7}{24}$ **b** $\frac{5}{12} \times \frac{9}{25}$

   c $16 \div \frac{8}{9}$ **d** $\frac{6}{7} \div \frac{15}{28}$

   e $2\frac{1}{3} \times 1\frac{1}{14}$ **f** $4\frac{3}{8} \div 1\frac{1}{4}$

3   Change these fractions into decimals using division. Give your answer as a recurring decimal if possible.

   a $\frac{7}{9}$ **b** $\frac{5}{11}$

   c $\frac{5}{12}$ **d** $\frac{6}{7}$

   e $\frac{7}{11}$ **f** $\frac{7}{13}$

4   Write these recurring decimals as fractions in their simplest form.

   a $0.313131...$ **b** $0.0\dot{6}$

5   Calculate these percentage changes.

   a Increase £90 by 4.5%

   b Decrease 1245 g by 61%

6   Jamie was paid £900 a month. His pay is increased to £960. What is the percentage increase?

7   A clock is reduced from £35 to £24.30 in a sale. What is the percentage reduction?

8   A laptop is reduced in price by 25% to £540. What was the original price?

9   In one year a tree grew 13% to a height of 1.92 m. What was its height at the beginning of the year?

10 £700 is placed in a savings account that pays 2.3% interest each year. How much is in the account at the end of three years? (assuming no money is removed and no additional money is paid in).

11 A car loses 14% of its value each year. What is it worth after 4 years if it was originally worth £13 000?

What next?

Score		
	0 – 4	Your knowledge of this topic is still developing. To improve look at Formative test: 3C-4; MyMaths: 1017, 1040, 1047, 1063, 1066, 1060 and 1073
	5 – 9	You are gaining a secure knowledge of this topic. To improve look at InvisiPen: 143, 144, 145, 152, 153, 154, 155 and 163
	10 – 11	You have mastered this topic. Well done, you are ready to progress!

1 Do these calculations leaving your answers as fractions in their simplest form.

a $1\frac{5}{6} + 2\frac{1}{8}$

b $3\frac{7}{9} - 1\frac{7}{8}$

c $2\frac{1}{5} \times 1\frac{5}{7}$

d $4\frac{7}{12} \div 1\frac{3}{8}$

e $5\frac{9}{14} + 2\frac{5}{21}$

f $3\frac{12}{35} - 1\frac{9}{10}$

g $2\frac{2}{7} \times 2\frac{5}{8}$

h $2\frac{8}{9} \div 2\frac{1}{6}$

2 Write these recurring decimals as fractions in their simplest form.

a 0.222...

b 0.1313...

c 0.732732...

d 0.45124512...

e 0.01666...

f 0.47777...

g 0.712929...

h 0.32323...

3 Find the reciprocals of each of these numbers without using a calculator.
Leave your answer in the most appropriate form.

a $\frac{1}{8}$

b $\frac{2}{7}$

c 1.4

d 16.4

4 Use an appropriate method of calculation to work out the following.

a Increase £35 by 35%

b Decrease £250 by 17.5%

c Increase 24.6 kg by 21%

d Decrease 176 km by 8.5%

5 a A CD costs £12.50. In a sale the price is reduced to £10.
What is the percentage reduction?

b A shirt normally costs £45. In a sale the price is reduced by 12%.
What is the sale price of the shirt?

6 a In a sale, a DVD is reduced in price by 15%. The sale price is £15.30.
What was the original price of the DVD?

b A cereal bar is increased in price by 8%. The new price is 81p.
What was the original price of the bar?

7 a Heidi bought a TV in a sale and saved £30. The label said that it was a 12%
reduction. What was the original price of the TV?

b At the start of December Keiley bought a pack of Christmas bells for £3.50.
On Christmas Eve she bought another pack but paid £1.40 less. What was the
percentage decrease in the price of the bells?

8 **a** A CD costs £11.50. In a sale the price is reduced by 35%.
What is the sale price of the CD?

b A scarf is reduced in price from £13 to £8.84.
What is the percentage reduction in price?

c An umbrella is reduced in price in a sale by 15%.
The new price of the umbrella is £11.05.
What was the original price of the umbrella?

d Jolene has increased in weight from 51 kg to 57 kg.
What is the percentage increase in her weight?

9 Work out the value of each of these items after the number of years stated.

Item	Cost	Time	Yearly percentage change
Car	£12 749	3 years	Decrease 11.5%
House	£299 999	25 years	Increase 3.4%
Loan	£6500	3 years	Increase 15.49%

10 The table shows the current prices of some items together with the yearly percentage change in their prices.

Item	Yearly percentage change	Price now
Mobile Phone	Decrease 17%	£29.99
House	Increase 7.1%	£315 000
Football Ticket	Increase 8.5%	£45

Calculate the price of each item

a 2 years ago **b** 5 years ago **c** 10 years ago.

11 Work out the value of each item after the number of years stated.

Item	Cost	Time	Percentage change each year
Car	£12 800	4 years	Decrease 12%
House	£220 000	3 years	Increase 4%
Savings	£1.5 million	5 years	2.49%

These questions test your knowledge of the topics in chapters 1-4. They give you practice in the types of questions that you may see in your GCSE exams.

There are 65 marks in total.

1 Estimate the answer to each of these calculations by first rounding the number to 1 significant figure.

    a $\dfrac{367 \times 6.8}{24.6}$ (1 mark)   **b** $\dfrac{961 \times 15.7}{46.9}$ (1 mark)

    c $\dfrac{6.9 \times 49.7}{26.4 \times 17.8}$ (1 mark)

2 Calculate the value of these, leaving your answer in standard form.

    a $(3 \times 10^{-4}) \times (4 \times 10^6)$ (1 mark)   **b** $(7 \times 10^{16}) \times (3 \times 10^{-1})$ (1 mark)

    c $(6 \times 10^{-34}) \times (5 \times 10^{14})$ (1 mark)   **d** $(2 \times 10^{-19}) \times (4 \times 10^2)$ (1 mark)

3 In an experiment the value of Young's Modulus, E, is given by

$$E = \frac{3.2 \times 10^3 \times 28}{4.15 \times 10^{-6} \times 0.103}$$

    a Estimate the value of Young's Modulus by rounding each number to 1 significant figure. (2 marks)

    b Use a calculator to determine the accurate value of Young's Modulus to 2 dp. (1 mark)

4 The distance from the Earth to the Sun is 93 million miles.
What is this distance in kilometres? (1 mark)

5 The Moon orbits the Earth approximately once every 28 days.
How long is this in seconds? (1 mark)

6 Which athletics race is longer, the 1500 m or the mile? (2 marks)

7 The formulae for the surface area and volume of a cone are

    Surface area = $\pi r^2 + \pi r s$ (1 mark)

    Volume = $\frac{1}{3}\pi r^2 h$ (2 marks)

    Show that the dimensions of each formula are correct.

8 The figure shows a trapezium prism.
Calculate the surface area of the prism. (5 marks)

9 A cylindrical metal rod is 14 cm long and 6 cm in diameter.
The mass of the rod is 2.5 kg.

    a Calculate the volume of the rod in cubic metres. (3 marks)

    b Determine the density of the rod. (2 marks)

    c A force of 3200 N is now exerted on the rod at its circular end.
Calculate the pressure on the rod. (3 marks)

10 Evaluate these quantities.

   a 9^{-2} (1 mark)   **b** $(m^4)^{-2}$ (1 mark)

   c $64^{\frac{1}{3}}$ (1 mark)   **d** $\dfrac{b^{\frac{2}{3}} \times b^{\frac{1}{3}}}{b^{-2}}$ (1 mark)

11 For these two shapes

   i write an expression for the surface area   (4 marks)

   ii expand the expression and simplify where possible.   (4 marks)

   a 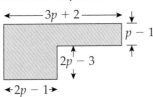   **b**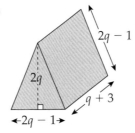

12 Factorise these expressions.

   a $24n^2 - 6n$ (1 mark)   **b** $16m^2 - 4$ (2 marks)

   c $6p^2 + 4p - 6$ (2 marks)

13 Make the letter in bold the subject of these formulae.

   a $E = \frac{1}{2}m\mathbf{v}^2$ (1 mark)   **b** $V = \frac{1}{3}\pi \mathbf{r}^2 h$ (2 marks)

14 A picture measures $16\frac{3}{8}$ inches long by $10\frac{3}{4}$ inches wide.

   a Calculate the perimeter of the picture.   (2 marks)

   b Calculate the area of the picture.   (2 marks)

   c The picture is now cropped taking $2\frac{1}{2}$ inches off the length and $1\frac{1}{8}$ inches

       off the width.

       What are the dimensions of the new picture?   (3 marks)

15 Write these recurring decimals as fractions.

   a 0.232323....   (2 marks)   **b** 0.533333...   (2 marks)

16 The cost of gas has increased by 22% during the last year so that the annual gas bill is
now £700. What was the cost of an annual bill before the increase?   (2 marks)

17 A customer invests £2000 in a savings account which accrues interest at 4.5% per year.
How much is in the account after

   a 1 year   (1 mark)   **b** 3 years?   (1 mark)

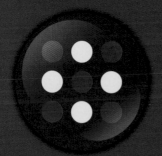

5 Angles and 2D shapes

Introduction

There are different units for measuring angles. The French tried to decimalise angles and invented a system where a right angle was equal to 100 grads. The most common measure is degrees, and there are 360 degrees in a circle. However, some mathematicians prefer a unit for measuring angles called the radian, and 1 radian = 57.2958 degrees!

What's the point?

The ability to estimate, measure and describe angles is highly important to all of us, whether we are designing stage lighting, creating a pattern for a dress or lobbing a ball towards a goal.

Objectives

By the end of this chapter, you will have learned how to ...

- Formulate a proof.
- Calculate and use the angle properties of polygons.
- Use the properties of a circle to calculate angles.
- Calculate an arc length and sector area of a circle.
- Recognise congruent shapes.

Check in

1 Calculate the missing angles.

a

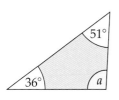

b

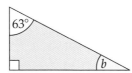

c

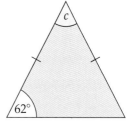

2 For these circles find the **i** perimeter **ii** area.

    a radius = 5 cm   **b** diameter = 12 m   **c** radius = 0.1 m

Starter problem

Inside a museum is a circular room.

The room will be lit by spotlights.

Each spotlight shines light in a beam

30° wide as shown in the diagram below.

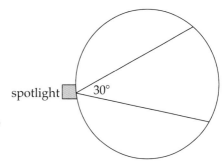

How many spotlights will be needed to completely

light the room and where should they be positioned?

Investigate using different spotlights with wider or

narrower beams of light.

5a Angle problems

The four interior angles fit together at a point.

360°

This diagram only shows that the **interior** angles of this particular quadrilateral add to 360° it is not a proof.

> A **proof** is a logical argument which proves a property for every situation.

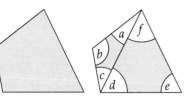

$a + b + c = 180°$ in one triangle
$d + e + f = 180°$ in the other triangle
$a + b + c + d + e + f$
$\qquad = 360°$ adding the equations

This proves that the interior angles of every quadrilateral add to 360°.

A demonstration only shows that a fact is true for one particular situation.

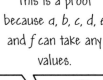

This is a proof because a, b, c, d, e and f can take any values.

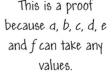

 Example

Use the diagram to prove the following statements.

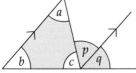

a The interior angles of every triangle add to 180°.
b The **exterior** angle of a triangle is equal to the sum of the opposite interior angles.

State which results you are using.

a
$\quad p = a$ Alternate angles
$\quad q = b$ Corresponding angles
$c + p + q = 180°$ Angles on a straight line
So $c + a + b = 180°$

b
$\quad p = a$ Alternate angles
$\quad q = b$ Corresponding angles
So $a + b = p + q$

Did you know?

Surveyors use a theodolite to measure Corresponding angles.

Exercise 5a

1 Calculate the unknown interior angles and state the type of triangle.

a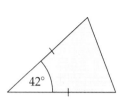

42°

b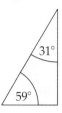

31°
59°

c

2 Calculate the values of the unknown angles.

a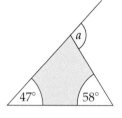

a
47° 58°

b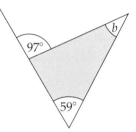

97°
b
59°

c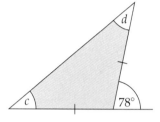

d
c 78°

3 Calculate the angles marked with letters.

a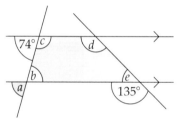

74° *c*
d
b
a *e*
135°

b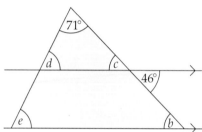

71°
d *c*
46°
e *b*

c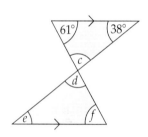

61° 38°
c
d
e *f*

Problem solving

4 Prove that the opposite angles of a parallelogram are equal.

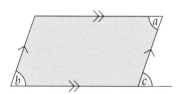

a
b
c

5 A triangle is drawn between two parallel lines.

  a Explain why

$$p + b + q = 180°.$$

  b Prove

$$a + b + c = 180°.$$

p *b* *q*
a *c*

6 Draw a quadrilateral and colour the angles.

Rotate the quadrilateral about the midpoint of each side to create a tessellation.

Show that the interior angles of a quadrilateral add to 360°.

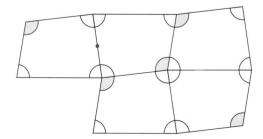

5b Angles in a polygon

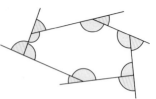

- The exterior angles of any **polygon** add to 360°.
- At each vertex, the exterior and the interior angles add to 180°.

You can split...

...a quadrilateral into two triangles.

Sum of interior angles
= 2 × 180° = 360°

...a pentagon into three triangles.

Sum of interior angles
= 3 × 180° = 540°

...a polygon with *n* sides into (*n* − 2) triangles.

- The interior angles of a polygon with *n* sides sum to (*n* − 2) × 180°.

Example

A polygon has 14 sides.

Calculate **a** the sum of the exterior angles   **b** the sum of the interior angles.

$\quad$ **a** $360°$ **b** $(14 - 2) \times 180° = 2160°$

- A **regular** polygon has equal sides and equal angles.
  ▶ A regular polygon with *n* sides has *n* lines of reflection **symmetry** and rotation symmetry of order *n*.

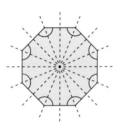

Example

A regular hexagon tessellates.
Explain why.

Calculate an interior angle.
Sum of the interior angles of a hexagon = (6 − 2) × 180°
$\qquad\qquad\qquad\qquad\qquad\qquad\quad = 720°$
$\qquad\qquad$ one interior angle = 720° ÷ 6 = 120°
3 × 120° = 360°
Angles at a point add to 360° so there are no gaps.
All sides are the same length so there are no overlaps.

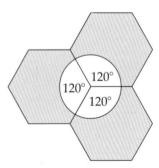

A **tessellation** is a tiling pattern with no gaps or overlaps.

Exercise 5b

1 Calculate the exterior and interior angle of
   a a regular pentagon
   b a regular decagon
   c a regular 20-sided polygon
   d a regular 30-sided polygon
   e a regular heptagon
   f a regular 17-sided polygon.

2 Calculate the unknown angles.

a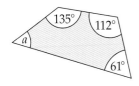

b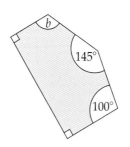

c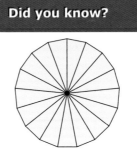

Problem solving

3 The exterior angle of a regular polygon is 15°.
   Calculate
   a the number of sides
   b the value of one interior angle
   c the sum of the interior angles
   d the number of lines of symmetry
   e the order of rotation symmetry of
   the polygon.

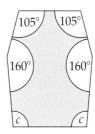
4 Eight kites fit together to form a regular octagon.
   Calculate
   a the four interior angles of the kite
   b the interior angle of a regular octagon.

5 A triangle, a regular hexagon and a regular dodecagon (12 sides) fit
   together as shown.
   a Use the angle properties of polygons to find each
   angle in the triangle and state its mathematical name.
   b Draw a diagram to illustrate the tessellation of these
   triangles, regular hexagons and regular dodecagons.

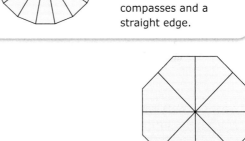

6 A polygon with n sides can be divided into $n - 2$ triangles.
   Each triangle has an angle sum of 180°.
   a Write an expression, in terms of n, for the sum of the interior angles.
   b What is the sum of the exterior angles?
   c Add your answers to find the sum of all the exterior and interior angles in a polygon.
   d Explain this answer.

5c Circle properties

A **circle** is a set of points equidistant from its centre O.
A **chord** is a straight line across the inside of a circle.
A **tangent** is a line that touches the outside of a circle.

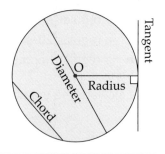

⬤ The tangent is perpendicular to the radius at the point
of contact.

⬤ A circle that passes through all the
vertices of a polygon is called a
circumscribed circle or a **circumcircle**.

⬤ An **inscribed** circle is the largest circle
that fits inside a polygon and touches
each side of the polygon.

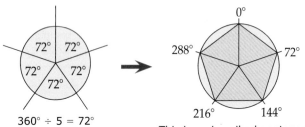

You can construct a regular polygon using a circumscribed circle.

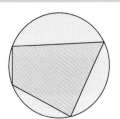

$360° \div 5 = 72°$
▲ Divide 360°
by the number
of sides of the
polygon.

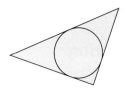

This is an inscribed pentagon.
▲ The chords and radii form
isosceles triangles to
create the polygon.

$2 \times 72 = 144$
$3 \times 72 = 216$
$4 \times 72 = 288$

Example

ABC is a triangle with a circumscribed circle, centre O.
Calculate the values of angles a, b, c and d.

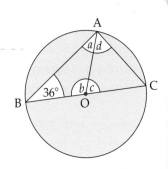

$a = 36°$ — AOB is an isosceles triangle
as OB = OA = radius of the circle.

$b = 180° - (36° + 36°)$ — Angle sum of a triangle.
$= 108°$

$c = 180° - 108°$ — Angle sum on a straight line.
$= 72°$

$d = (180° - 72°) \div 2$ — AOC is an isosceles triangle
$= 54°$ — as OA = OC = radius of the circle.

6

Exercise 5c

1 O is the centre of the circle and AB is the tangent at Q.

    a State the value of angle OQA.

    b Explain why triangle OPQ is isosceles.

    c Calculate the values of angles *a* and *b*.

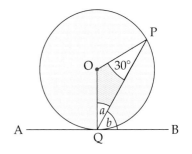

2 A circumscribed circle of a square
has a radius of 10 cm.
Calculate the area of the square.

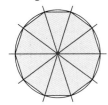

3 **a** Construct an inscribed regular decagon,
using a circle of radius 5 cm.

    b Measure the interior angle of the decagon.

    c Calculate the sum of the interior angles.

    d Check your answer using the formula:

        sum of the interior angles of a polygon = $(n - 2) \times 180°$

        where *n* is the number of sides.

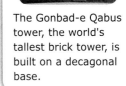
4 BC is the diameter of a circle with centre O.
A is any point on the circumference.

    a Explain why angle ABO = *b*.

    b Explain why angle ACO = *c*.

    c Find the angle sum of triangle ABC in terms of *b* and *c*.

    d Find the value of *b* + *c*.

    e What is the value of angle BAC?

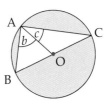

Problem solving

5 **a** Use the circumscribed circle of a regular
pentagon to draw a 5-pointed star.

    b Calculate

      i the angle at a point of the star

      ii the sum of the angles at all 5 points.

    c Repeat for a 6-pointed and an 8-pointed star.

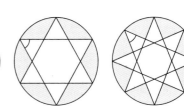

You can calculate the **circumference** and area of a circle using these formulae.

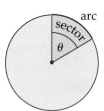

● Circumference = π × **diameter**	$C = \pi d$
● Circumference = π × 2 × **radius**	$C = 2\pi r$
● Area of a circle = π × radius × radius	$\text{Area} = \pi r^2$
where π = 3.141592..., $d = 2r$	

C = circumference
d = diameter
r = radius

Example

Calculate the radius of the circle that has an area of 50 cm². Use π = 3.14, and give your answer to 1 dp.

Area $= \pi r^2$
$50 = \pi r^2$
$r^2 = 50 \div \pi$
$r^2 = 15.92$
$r = \sqrt{15.92} = 3.99$
$r = 4.0$ cm (to 1 dp)

Area = 50 cm²

The length of an **arc** is a fraction of the circumference.
The fraction is proportional to the value of the angle θ.

● Arc length $= \dfrac{\theta}{360} \times$ circumference
● Area of **sector** $= \dfrac{\theta}{360} \times$ area of the circle.

▲ The Greek letter θ (theta) is often used for angles.

Example

Calculate the arc length and area of the sector shaded pink.
Give your answers accurate to 1 dp.

radius = 10 cm

Arc length $= \dfrac{\theta}{360°} \times$ circumference

$= \dfrac{60°}{360°} \times 2 \times \pi \times 10$

$= 10.5$ cm (to 1 dp)

Area of sector $= \dfrac{\theta}{360°} \times$ area of the circle

$= \dfrac{60°}{360°} \times \pi \times 10 \times 10$

$= 52.3$ cm² (to 1 dp)

▲ A dart board has 20 sectors.

7

Exercise 5d

Use π = 3.14 for all the questions on this page and give your answers accurate to 1 dp.

d is the diameter
r is the radius

1 Calculate the circumference and area of each of these circles.

   a *r* = 7.5 cm   **b** *d* = 11 cm   **c** *r* = 25 mm   **d** *d* = 30 m

2 Calculate the arc length and area of each pink and blue sector.

   a

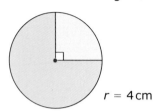

r = 4 cm

   b

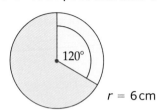

120° *r* = 6 cm

   c

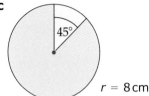

45° *r* = 8 cm

   d

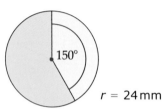

150° *r* = 24 mm

   e

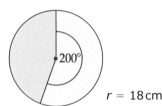

200° *r* = 18 cm

   f

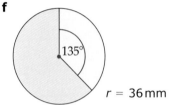

135° *r* = 36 mm

Problem solving

3 The circumference of a bicycle wheel is 216 cm. Calculate
   a the diameter and radius of the wheel
   b the number of turns of the wheel in a journey of one kilometre.

4 The area of the top of the the circular pie is 110 cm².
Calculate the radius of the pie.

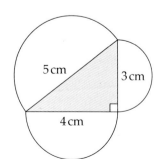

5 The length of the minute hand on a clock is 10 cm.
The length of the hour hand is 5 cm.
Calculate the distance the tip of each hand travels in
   a one hour   **b** one day   **c** one year.

6 A semicircle is drawn on each side of a
3 cm, 4 cm, 5 cm right-angled triangle.
   a Calculate the area of each semicircle.
   b Add the areas of the two smaller semicircles
      and compare the answer with the area of the
      semicircle on the hypotenuse.
   c Repeat this with other right-angled triangles.
   d Use algebra to explain this rule.

5 cm   3 cm   4 cm

○ **Congruent** objects are exactly the same shape and size.

If shapes are congruent, then
corresponding angles and
corresponding sides are equal.

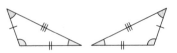

▲ A person and their image are identical but reversed.

Two triangles are congruent if they satisfy one of these four sets of conditions.

▲ two sides and the included angle are equal (SAS)   ▲ two angles and the included side are equal (ASA)

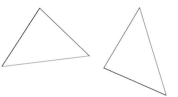

▲ three sides are equal (SSS)   ▲ a right angle, hypotenuse and side are equal (RHS)

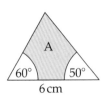

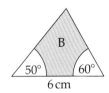

60°   A   50°   50°   B   60°
   6 cm   6 cm

> Reflect triangle B and then the triangles will fit exactly on top of each other.

The two angles and the included side of these triangles are equal (ASA) so the triangles are congruent.

○ You can use **congruence** to prove mathematical properties of shapes.

Example

ABC is a triangle with AB = AC.
AX is perpendicular to BC.
Prove angle ABX = angle ACX and BX = CX.

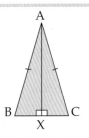

AB = AC   and   angle BXA = angle CXA = 90°
AX = AX   It is the same line.
Therefore triangles ABX and ACX are congruent by RHS.
This means the corresponding angles ABX and ACX are equal.
The base angles of an isosceles triangle are equal.

This means the corresponding sides BX and CX are equal.
The foot of the perpendicular bisects the base of an isosceles triangle.

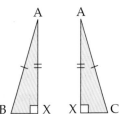

8

Exercise 5e

1 Match the congruent triangles.

Use SAS, ASA, SSS or RHS to help you decide.

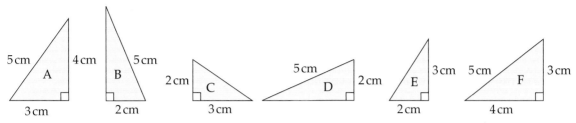

5 cm 4 cm 5 cm 2 cm 5 cm 2 cm 3 cm 5 cm 3 cm

A B C D E F

3 cm 2 cm 3 cm 2 cm 4 cm

2 Name the two congruent triangles.
Explain your answer using SAS,
ASA, SSS or RHS.

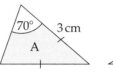

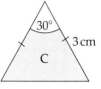

70° 3 cm   75°   30° 3 cm

A   B   C

3 cm

Problem solving

3 Explain why these two triangles are congruent.

Hence find *x* and *y*.

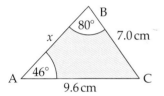
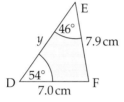

B
80° 7.0 cm
x
A 46° C
9.6 cm

E
46° 7.9 cm
y
D 54° F
7.0 cm

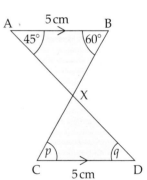

A 5 cm B
45° 60°
X
p q
C 5 cm D

4 In the diagram on the right, AB and CD are two parallel lines.
AB = CD = 5 cm.

a Find the values of angles *p* and *q*.

b Explain why triangles ABX and DCX are congruent.

5 These two diagrams show constructions for the angle bisector
of angle ABC and perpendicular bisector of line PQ.

a

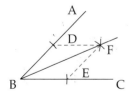

A
D
F
B E
C

b

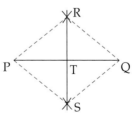

R
P T Q
S

i Explain why BD = BE and DF = EF.

ii What shape is BDFE?

iii Show that triangles BDF and BEF are
congruent; hence deduce that angle
DBF = angle EBF.

i Show that triangles PRS and QRS
are congruent and identify all pairs
of equal angles.

ii Show that triangles PRT and QRT are
congruent; hence deduce that PT = QT
and angle PTR = angle QTR = 90°.

Check out

You should now be able to ...

Test it ➡
Questions

✓ Calculate and use the angle properties of polygons.	6	1 – 3
✓ Use the properties of a circle to calculate angles.	6	4
✓ Calculate an arc length and sector area of a circle.	7	5, 6
✓ Recognise congruent shapes.	8	7

Language   Meaning

Example

Language	Meaning	Example
Proof	A set of mathematical statements that proof a rule is true in all cases.	$2n$ is always even for all integer values of n so $2n + 1$ must always be odd
Circle	A set of points equidistant from the centre O.	Tangent, Diameter, O, Radius, Chord
Chord	A straight line across the circle.	
Tangent	A line the touches the outside of a circle.	
Circumference	The edge of a circle.	
Arc	Part of the circumference.	arc, sector, θ
Sector	A region of the circle enclosed by two radii and an arc.	
Congruent **Congruence**	Two shapes are congruent when they are the same shape and size.	

1 Calculate the value of the angles denoted by letters. Explain which geometric facts you use.

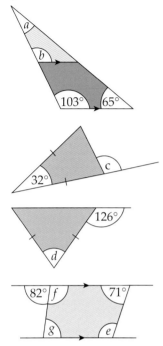

2 Calculate the exterior and interior angle of
  a a regular hexagon
  b a regular 18-sided polygon.

3 The exterior angle of a regular polygon is 8°. Calculate
  a the number of sides
  b the size of one interior angle
  c the sum of the interior angles
  d the number of lines of symmetry
  e the order of rotation symmetry of the polygon.

4 C is the centre of the circle and XY is the tangent at A.

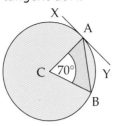

Calculate the size of angles
  a CAB   **b** ABC   **c** BAY

5 For this circle, calculate
  a the arc length **b** the area of the sector
Give your answers as fractions in terms of π.

6 The area of a semi-circle is $95\,cm^2$. Calculate its radius. Give your answer to 2 dp.

7

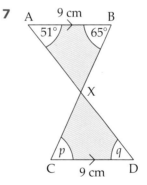

  a Calculate angles p and q.
  b Is triangle ABX similar or congruent to triangle CXD?

What next?

1 – 3	■	Your knowledge of this topic is still developing. To improve look at Formative test: 1002, 1009, 1100, 1118, 1148 and 1320
4 – 6	▨	You are gaining a secure knowledge of this topic. To improve look at InvisiPen: 317, 346, 347, 348 and 354
7	▨	You have mastered this topic. Well done, you are ready to progress!

Score

1 Calculate the values of the marked angles.

a

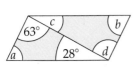
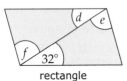

parallelogram

b

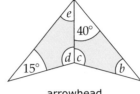

arrowhead

c

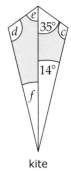

kite

d

rectangle

e

isosceles trapezium

2 Calculate the exterior and interior angle of

   a a regular hexagon   **b** a regular nonagon (9 sides)

   c a regular 18-sided polygon   **d** a regular octagon.

3 For each shape in question **2**, explain if the shape will tessellate.

4 The interior angle of a regular polygon is 150°.

   a Find the exterior angle.

   b How many sides does the polygon have?

5 You can draw a circumscribed circle for a square.
If possible, draw a circumscribed circle for

   a a rectangle   **b** a kite   **c** a rhombus

   d a parallelogram   **e** an isosceles trapezium.

6 Calculate the values of angles *a* and *b*.

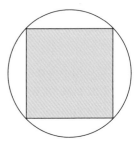

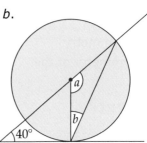

7 Calculate the arc lengths and the areas of the shaded sectors.
Round your answers to an appropriate degree of accuracy.

a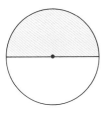

$r = 9\,cm$

b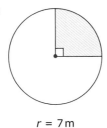

$r = 7\,m$

c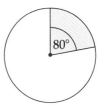

$r = 2.5\,cm$

d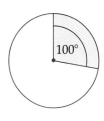

$r = 3.5\,cm$

8 Calculate the shaded area.

a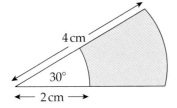

b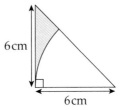

9 **a** Find the value of a.

b Explain why the two triangles are congruent using SSS, SAS, ASA or RHS.

c Explain why these two triangles are congruent.
Hence find x and y.

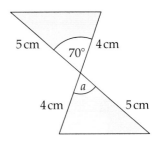

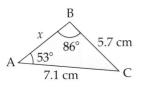 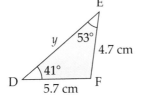

10 Which pair of these triangles is congruent?
Explain your reasons.

A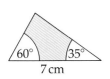

B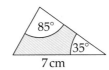

C

D

6 Graphs

Introduction

Mobile phone companies offer the latest phones with all the latest features to encourage you to spend your money with them. When you are choosing your next mobile phone it pays not just to look at the features on the phone but also to compare the price plans on offer. The price plans are often hard to compare and the best way is to represent the different set of information as graphs.

What's the point?

In business, many problems are about obtaining the best results within a given set of constraints. Answers are not often clear cut, but the plotting of linear functions lies at the heart of the decision making process.

Objectives

By the end of this chapter, you will have learned how to ...

- Plot graphs of linear functions and find gradients.
- Find the equation of straight-line graphs.
- Recognise and plot graphs of quadratic functions.
- Recognise and plot graphs of cubic functions.
- Plot and interpret distance-time graphs.
- Plot and interpret graphs from other real-life situations.
- Identify trends in time-series graphs and data.
- Read and interpret exponential and reciprocal graphs.

1 $y = 6 - 3x$

Find y when

   a $x = 1$ **b** $x = 2$ **c** $x = 4$ **d** $x = -3$

2 Plot the following line graphs, each on a separate grid.

   a $y = 2x + 1$ **b** $y = 5 - 3x$ **c** $x + y = 12$ **d** $x + 3y = 10$

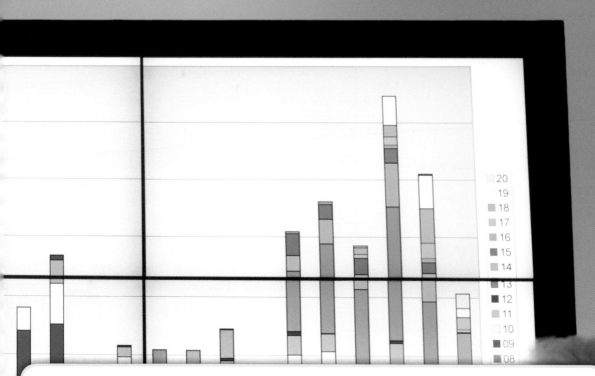

Starter problem

Here are some functions.

$y = 3x + 2$ $y = x^2 + 3x + 2$ $y = x^3 - 3x^2 - x + 3$

$y = \dfrac{3}{x}$ $y = 3^x$

Draw graphs of each function using x values from -4 to +4.

What is the same and what is different about these graphs?

6a The gradient of a straight line

● The **gradient**, m, is a measure of steepness.

$$m = \frac{\text{amount you go up}}{\text{amount you go across}} \quad \text{or} \quad \frac{\text{up}}{\text{across}}$$

A gradient of $\frac{1}{2}$ represents 1 up and 2 across.

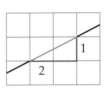

● Lines which slope upwards ⟋ have a positive gradient.
● Lines which slope downwards ⟍ have a negative gradient.

▲ You may see the gradients of roads given as percentages.

Example

Find the gradient of the line joining the point (1, 3) to the point (4, 5).

Draw a diagram and count squares.
The **line segment** travels
3 squares across and 2 up.

Gradient $= \frac{2}{3}$

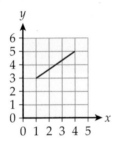

● The gradient can also be found using a formula.

For (x_1, y_1) to (x_2, y_2), $m = \dfrac{y_2 - y_1}{x_2 - x_1}$

For example, (1, 3) to (4, 5), $m = \dfrac{5 - 3}{4 - 1} = \dfrac{2}{3}$

Example

Find the gradient of the line joining the points (0, 6) to (4, -2).

$(x_1, y_1) = (0, 6)$ and $(x_2, y_2) = (4, -2)$

The gradient, $m = \dfrac{y_2 - y_1}{x_2 - x_1} = \dfrac{(-2) - 6}{4 - 0}$

$\qquad\qquad = \dfrac{-8}{4}$

$\qquad\qquad = -2$

Check the sign of the gradient. Imagine a line sloping from (0, 6) to (4, -2). The line slopes downwards so the gradient is negative.

Exercise 6a

1 Match the gradients with the correct line segments.

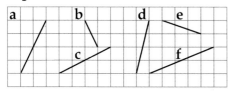

A -2	**B** 4		**C** 2
D $\frac{2}{5}$	**E** $-\frac{1}{3}$		**F** $\frac{1}{2}$

2 **a** Find the gradient of the line segment joining each of the following pairs of points.

     i (4, 7) to (5, 13)

2   **ii** (0, 9) to (2, 10)

     iii (-2, 8) to (2, 12)

     iv (-3, 6) to (-6, 9)

     v (-4, -7) to (-1, 10)

   b Put your answers in ascending order of steepness.

3 Match the line segments with the same gradient.

   a (6, -10) to (9, -9)

   b (-4, 10) to (-2, 6)

   c (100, 7) to (103, 1)

   d (1, 1) to (2, 4)

   e (0, 0) to (3, 1)

   f (10, 4) to (14, 16)

Problem solving

4 Four students found the gradient of the line joining (-2, 8) to (10, 2). Only one student did so correctly. Find the correct answer and explain why the others were incorrect.

Anil $m = \dfrac{10 - (-2)}{2 - 8}$

Beth $m = \dfrac{2 - 8}{(-2) - 10}$

Cheryl $m = \dfrac{2 - 8}{10 - (-2)}$

Dev $m = \dfrac{2 - 8}{10 - 2}$

5 Write an expression for the gradient of the line joining (a, b) to (c, d).

6 Find the value of h if the gradient joining the points $(h, 4h)$ to $(3h, 30)$ is 1.

7 A quadrilateral ABCD is formed by joining A (2, 12), B (5, 27), C (3, 28) and D (-1, 8). Find the gradient of each of its edges. What type of quadrilateral is it?

Did you know?

The gradient of a flight of stairs is given by the rise divided by the tread. Building regulations place limits on their values. By law the gradient of new stairs must be less than 0.9.

It usually helps to draw a diagram.

6b Graphs of linear functions

The equation of a **straight line graph** can be expressed in the form
$y = mx + c$
where m is the gradient and c is the y-intercept.

A straight line has equation $6x - 2y = 1$
Write   **a**   the gradient
         b   the coordinates of the y-intercept.

The y-intercept is the point at which the line cuts the y-axis. The coordinates of this point are $(0, c)$.

Rearrange into the standard form.

$6x - 2y = 1$

$\qquad 6x = 1 + 2y$ Add $2y$ to both sides.

$6x - 1 = 2y$ Subtract 1 from both sides.

$3x - \dfrac{1}{2} = y$ Divide both sides by 2.

$\qquad y = 3x - \dfrac{1}{2}$ Compare this equation with $y = mx + c$.

a   Gradient, 3   **b**   Intercept, $\left(0, -\dfrac{1}{2}\right)$

A linear function contains two variables but with no powers.
$y = 5x + 2$ is a linear function.
$y = x^2 + x - 12$ is not a linear function.

To plot the graph of a **linear function**, construct a table of values.
▶ A linear function will always produce a straight line graph.

a   Plot the graph of $y = 5 - 3x$ on a pair of coordinate axes.
b   Without plotting, explain how you know that the graphs
$y = 5 - 3x$ and $y = 2 - 3x$ do not intersect.

A gradient of -3 means that for every 1 unit across you move 3 units *down*.

a

x	0	1	2
y	5	2	-1

b $y = 5 - 3x$ and $y = 2 - 3x$
both have gradient $= -3$
and so they are parallel.
Parallel lines do not intersect.

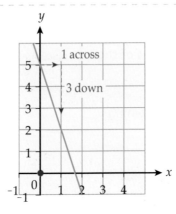

Exercise 6b

1 Match each graph with its equation.

a 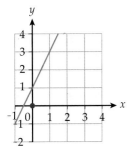 **b** 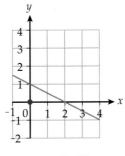 **c**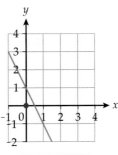

A $y + 2x = 1$ **B** $y = 2x + 1$ **C** $y = 1 - \frac{1}{2}x$

2 Construct a table of values for each of these equations and hence plot their graphs.

a $y = 3$ **b** $x = -1$ **c** $y = 5x + 1$ **d** $2y = x + 2$ **e** $2x + y = 10$

3 **a** Copy and complete the table of values for the equation $y = 4 - 3x$.

x	0	1	2
y		1	

b Plot these values on a set of coordinate axes.

c Write the equation of a line that is parallel to $y = 4 - 3x$.

Problem solving

4 Are these statements true or false?
   a The point (2, 5) lies on the line $y = 4x - 3$.
   b The lines $2y = 6x + 1$ and $y - 3x = 1$ do not intersect.
   c The lines $x = 2$ and $y = -3$ are perpendicular.

5 Write the equations of these straight lines.
   a The line has a gradient of 3 and passes through (0, 5).
   b The line has a gradient of 2 and passes through (3, 5).
   c The line passes through (0, 1) and (1, 5).

6 Choosing from these equations find
   a a pair of parallel lines.
   b a pair of lines with the same y-intercept.
   c a pair of perpendicular lines.

A $y = 2x - 1$ **B** $y = x + 2$ **C** $x + y = 5$

D $y - 2x = 3$ **E** $y = 2(1 - x)$

A sketch will help you decide which lines are perpendicular.

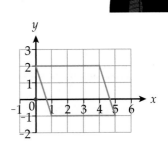

7 Write the equations of the four straight lines that make up the sides of this parallelogram.
   What do you notice about the equations?

The opposite sides of a square are **parallel**. Their gradients are equal.

The adjacent sides of a square are **perpendicular**. They meet at right angles. Their gradients are of opposite signs – one is positive and the other is negative.

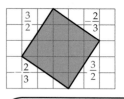

- Parallel lines have the same gradient.

- Perpendicular lines have gradients which are the negative **reciprocal** of one another.
  - ▶ For perpendicular lines, the **product** of their gradients is -1.

$$\frac{3}{2} \times \left(-\frac{2}{3}\right) = -\frac{6}{6} = -1$$

In general, lines with gradient m and $-\frac{1}{m}$ are perpendicular.

> Think of a reciprocal as "flipping" a fraction. For example, $\frac{1}{2}$ and $\frac{2}{1}$ (= 2), or $\frac{3}{4}$ and $\frac{4}{3}$ are reciprocal pairs.

Example

The equations of three lines are

$$2y = 8x + 5 \qquad y = 1 - \frac{1}{4}x \qquad x + 4y = 3.$$

Find the two lines that are parallel and the line that is perpendicular to them.

Rewrite the equations in the form $y = mx + c$.

$2y = 8x + 5 \qquad \Rightarrow y = 4x + 2.5$ This line has gradient 4.

$y = 1 - \frac{1}{4}x \qquad \Rightarrow y = -\frac{1}{4}x + 1$ This line has gradient $-\frac{1}{4}$.

$x + 4y = 3 \qquad \Rightarrow 4y = -x + 3$

$\qquad\qquad\qquad \Rightarrow y = -\frac{1}{4}x + \frac{3}{4}$ This line has gradient $-\frac{1}{4}$.

The lines $y = 1 - \frac{1}{4}x$ and $x + 4y = 3$ have the same gradient, $-\frac{1}{4}$. They are parallel.

The line $2y = 8x + 5$ has gradient 4 and $4 \times \frac{-1}{4} = -1$, so it is perpendicular to the other two lines.

Example

A line is parallel to $y = \frac{1}{3}x + 1$ and passes through (3, 5).
Write the equation of this line.

$y = \frac{1}{3}x + 1$ This line has gradient $\frac{1}{3}$.

$y = \frac{1}{3}x + c$ A parallel line will also have gradient $\frac{1}{3}$.

$5 = \frac{1}{3} \times 3 + c$ Substitute (3, 5) into this equation.

$5 = 1 + c \qquad \Rightarrow \qquad c = 4$

The equation is $y = \frac{1}{3}x + 4$

Exercise 6c

1 Copy and complete this table of gradients.

	Gradient of line	Gradient of parallel line	Gradient of perpendicular line
	5		
	$-\frac{1}{3}$		
Equation of line	$\frac{3}{4}$		
	1.5		
$y = 4x + 3$			
$3y = 8x + 1$			

2 Are these statements true or false?

a The line $y = 5x + 4$ has a gradient of 4.

b The lines $y = 5x + 4$ and $2y = 5x + 7$ are parallel.

c The lines $y = \frac{1}{4}x$ and $y = 4x$ are perpendicular.

d Lines with gradient 0.125 and $\frac{1}{8}$ are parallel.

e Lines with gradient $-1\frac{1}{3}$ and $\frac{3}{4}$ are perpendicular.

Problem solving

3 Given the equation of one line, find the equation of the other line.

a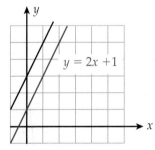

b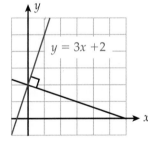

c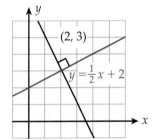

4 A triangle is formed by joining the points A (1, 1), B (3, 11) and C (6, 0). By finding the gradient of each side of the triangle, decide if ABC is a right-angled triangle.

Draw a diagram to help you.

5 A rectangle is drawn on coordinate axes. One edge has equation $y = 2x + 4$. Give possible equations for the other three sides.

6 Explain why perpendicular lines must have gradients that are the negative reciprocals of one another.

▶ Explain why one gradient must be positive and one must be negative.

▶ Then explain why the gradients cannot have the same numerical value.

You may find diagrams on squared paper useful to support your explanations.

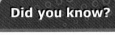 **MyMaths.co.uk**　　Q 1314　　SEARCH

6d Quadratic graphs 1

A **quadratic function** has an x^2 term as its highest power of x.

The curved graph produced by a quadratic function is called a **parabola**.

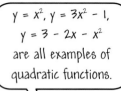

$y = x^2$, $y = 3x^2 - 1$,
$y = 3 - 2x - x^2$
are all examples of quadratic functions.

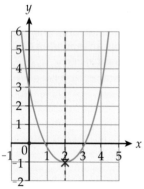

A symmetrical · shaped graph is produced by positive x^2 terms.

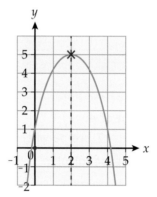

A symmetrical · shaped graph is produced by negative x^2 terms.

Parabolas have a line of symmetry that passes through the minimum or maximum point.

Example

a Plot the graph of the function $y = x^2 - 2x - 3$.

b Write the equation of the line of symmetry of the curve.

c Write the coordinates of the minimum point of the curve.

a Construct a table of values.

x	-2	-1	0	1	2	3	4
x^2	4	1	0	1	4	9	16
$-2x$	4	2	0	-2	-4	-6	-8
-3	-3	-3	-3	-3	-3	-3	-3
y	5	0	-3	-4	-3	0	5

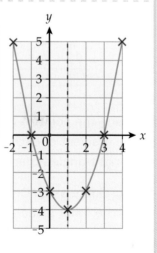

Plot the coordinate pairs.

(-2, 5), (-1, 0), (0, -3), (1, -4), (2, -3), (3, 0) and (4, 5)

This curve will be a · shaped parabola as the x^2 term is positive.

b The line of symmetry is $x = 1$

c The minimum point is (1, -4).

Exercise 6d

1 a Copy and complete the table of values for the function $y = x^2 - x$.

x	-3	-2	-1	0	1	2	3	4
x²		4					9	
-x		2					-3	
y		6					6	

b Plot the graph of $y = x^2 - x$

c Write the equation of the line of symmetry of the curve.

You will have to use the graph to find the coordinates of the minimum point.

2 a Copy and complete the table of values for the function $y = x^2 + x - 2$.

x	-4	-3	-2	-1	0	1	2	3
x²		9					4	
x		-3					2	
-2		-2					-2	
y		4					4	

b Plot the graph of $y = x^2 + x - 2$

c Write the coordinates of the minimum point of the curve.

Problem solving

3 a Plot the graph of $y = x^2 - 4x + 3$ for values of x from -1 to 5.

b Write the equation of the line of symmetry of the curve.

c Write the coordinates of the y-intercept of the curve.

d Write the coordinates of the points where the curve crosses the x-axis.

4 This is the graph of $y = x^2 - x - 6$

a Make an accurate copy of this graph on squared paper.

b On the same diagram, plot the graph of the linear function $y = x - 3$

c Write the coordinates of the points of intersection of these two graphs.

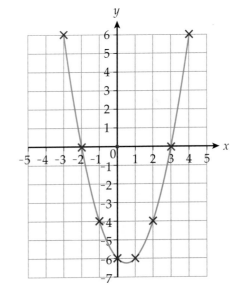

5 Look again at the graph is question 4.

a Write down the coordinates of the points where the quadratic graph intersects the x-axis.

b Hence solve the equation $x^2 - x - 6 = 0$.

c How can you use the graph to solve $x^2 - x - 4 = 0$?

● If the coefficient of x^2 is positive the parabola has a **minimum** point.

● If the coefficient of x^2 is negative the parabola has a **maximum** point.

Did you know?

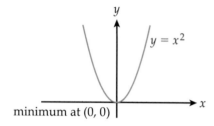

$y = x^2$

minimum at (0, 0)

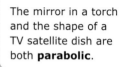

maximum at (0, 0)

$y = -x^2$

The mirror in a torch and the shape of a TV satellite dish are both **parabolic**.

Graphs of quadratic functions can after describe real-life situations.

Example

A goalkeeper kicks a football from his area. The height of the football at any point is given by the equation

$y = 6x - x^2$

where

> y is the height of the ball above the ground (m)
> x is the time from when it was kicked (s).

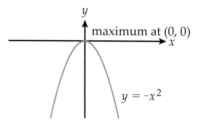

9m

6m

a Plot a graph to show the path of the football for values of x from 0 to 6.

b Use this to work out the highest point above the ground that the ball reached.

a

x	0	1	2	3	4	5	6
$6x$	0	6	12	18	24	30	36
$-x^2$	0	-1	-4	-9	-16	-25	-36
y	0	5	8	9	8	5	0

Plot the points (0, 0), (1, 5), (2, 8) etc. and join them with a smooth curve.

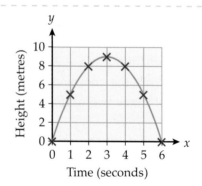

Height (metres) / Time (seconds)

b The football reached a maximum height of 9 m when $x = 3$ s.

Exercise 6e

1   For each equation given

    i   use a completed table of values to plot a smooth parabola

    ii decide if your parabola has a maximum or minimum point
and give its coordinates

    iii state the equation of the line of symmetry.

a $y = 2x^2 - 8x + 3$

x	-1	0	1	2	3	4	5
$2x^2$					18		
-8x					-24		
+3	3	3	3	3	3	3	3
y					-3		

b $y = 6x - x^2$

x	-1	0	1	2	3	4	5	6
6x								
$-x^2$								
y								

Did you know?

When a uniform load is hung from the cables of a suspension bridge they make the shape of a parabola.

Problem solving

2   John throws a ball in the air. Its height, y metres, after x seconds is given by the equation $y = 16x - 4x^2$.

  a Plot a graph to show the path of the ball over 5 seconds.

  b Find the coordinate of the highest point of your graph.
What does this mean in real life?

  c When does the ball hit the ground?

  d Find y when $x = 5$. Explain why your result is impossible in real life.

3   A rectangular conservatory is built on a house using 50 metres of framework.

Let x be the width of the conservatory.

  a In terms of x, write expressions for the conservatory's

    i length   **ii** area, y.

  b Plot a graph of area y against width x.

  c Use your graph to find the largest possible area of this conservatory
and explain why, in real life, a homeowner may wish to maximise the area.

House — Framework, x

4   Try to find the equations of the parabolas used in one or more suspension bridges from around the world.

Can you use your findings to form an accurate scale drawing of your bridge?

Can you find out some other real-life uses of quadratic graphs?

MyMaths.co.uk   Q 1316   **SEARCH**

6f Cubic graphs

⬤ **Cubic** expressions contain x^3 as the highest power.
▶ Cubics have a characteristic S-shape when plotted.

The simplest cubic function is $y = x^3$

x	-3	-2	-1	0	1	2	3
y	-27	-8	-1	0	1	8	27

When $x = -2$
$$y = (-2)^3$$
$$= (-2) \times (-2) \times (-2)$$
$$= -8$$

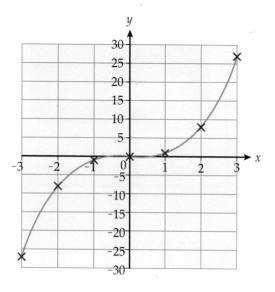

Plot the points $(-3, -27)$, $(-2, -8)$, $(-1, -1)$, etc. and join them with a smooth curve. Don`t use a ruler to join the points.

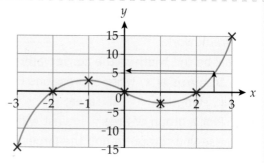

Example

Plot the graph $y = x^3 - 4x$ for values of x in the range $-3 \leq x \leq 3$
Use your graph to estimate the value of $2.5^3 - 4 \times 2.5$

Construct a table of values.

x	-3	-2	-1	0	1	2	3
x^2	-27	-8	-1	0	1	8	27
$-4x$	12	8	4	0	-4	-8	-12
y	-15	0	-3	0	-3	0	15

Read from the graph.
$y \approx 5.5$

To find $2.5^3 - 4 \times 2.5$, find 2.5 on the x-axis. Draw a straight line up to the graph and go across to the y-axis.

Exercise 6f

1 Match each equation with its graph.

a $y = x^2 + 1$ **b** $y = 1$ **c** $y = x + 1$

d $y = 1 - x^2$ **e** $y = x^3 + 1$ **f** $x = 1$

A ——   B ⋃   C ╱   D ∫   E │   F ⋂

2 Copy and complete the table of values for each cubic function and use your values to plot the graph of the function.

Think carefully about the choice of scale for each graph.

a $y = x^3 + x + 2$

x	-3	-2	-1	0	1	2	3
x³			-1				
x			-1				
2	2	2	2	2	2	2	2
y				2			

b $y = -x^3$

x	-3	-2	-1	0	1	2	3
y						-8	

c $y = x^3 - 3x^2 + 9$

x	-3	-2	-1	0	1	2	3
x³						8	
-3x²						-12	
9						9	
y						5	

$-x^3$ means first cube the number and then multiply by -1.

Problem solving

3 a Plot the graph of $y = x^3 - x + 4$ for $-3 \le x \le 3$.

   b On the same pair of axes, plot the graph of $y = x + 5$.

   c At which points do your graphs intersect?

   d Find the equation of a straight line that intersects the curve $y = x^3 - x + 4$

     i only once   **ii** three times.

4 The graph of $y = x^2$ is a parabola and the graph of $y = x^3$ is an S-shaped curve. Using graphical software or otherwise, investigate the shape of $y = x^4$, $y = x^5$, etc. What do you notice? Can you explain why this happens?

Did you know?

Cubic equations are often used by engineers to design the curves of many objects including roller coasters.

⬤ **Distance-time graphs** are used to represent journeys.

▶ The **gradient** of a distance-time graph is the speed.

$$\text{Speed} = \frac{\text{Distance covered}}{\text{Time taken}}$$

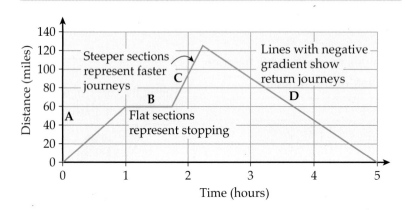

This distance-time graph represents the journey of a police car.

A The police car left the station and travelled at a **speed** of 60 miles per hour for one hour. ➡ To go to the scene of a crime.

B The car was **stationary** for 45 minutes. ➡ Whilst they observed the suspects.

C The car then travelled 65 miles in half an hour. Its speed was 130 miles per hour. ➡ As they chased the suspects.

D The police car then returned to the station.

At what speed did the police car return to the police station?

Use the gradient formula on the line joining $(2\frac{1}{4}, 125)$ to $(5, 0)$.

$$\text{Speed} = \frac{y_2 - y_1}{x_2 - x_1} = \frac{0 - 125}{5 - 2\frac{1}{4}} = \frac{-125}{\frac{11}{4}}$$

$$= -125 \times \frac{4}{11} = -45\frac{5}{11}$$

The car returned at $45\frac{5}{11}$ miles per hour.

Take care with time as it is not decimal. $2\frac{1}{4}$ hours is not the same as 2 hours 25 minutes.

Describe the speed of a car in this distance-time graph.

The gradient of the curve gradually gets steeper so the speed is gradually increasing. The graph represents **acceleration**.

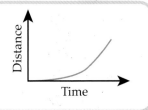

Exercise 6g

1 Match each distance-time graph sketch with a description.

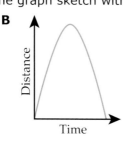 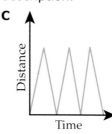 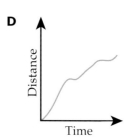

A **B** **C** **D**

a A ball thrown up into the air.

b A man walking to and from the local shop.

c A car stuck in traffic.

d A swimmer doing lengths of a pool.

Problem solving

2 The distance-time graph shows the movements of a rabbit.

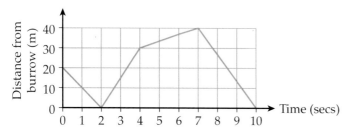 

a How do you know that the rabbit did not begin its
   journey at its burrow?

b Did the rabbit rest at any point during its journey?

c How often did the rabbit visit its burrow?

d What was the rabbit's speed during
   i   the first second?   ii   the last second of its journey?

e What was the rabbit's average speed during the entire journey?

3 a Make a copy of the graph from question **2**.

   b Add to this graph a second rabbit making the following journey.

   > The rabbit leaves the burrow and travels at a speed of 5 m/s for three seconds, then rests for one
   > second. After a further two seconds, the rabbit is 40 m from the burrow. The rabbit then returns
   > to the burrow at a speed of 17 m/s.

4 Make up your own journey story about an object of your choice
   and show the information on a distance-time graph.

5 For each description given in question **1**, draw a *speed*-time graph.

Many real-life situations can be represented by a graph.

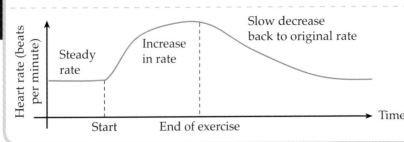

Height of female students at St Anne's College

A **sketch** graph is not accurate but shows key features clearly.
A **plot** is an accurate graph.

This **sketch** shows that most students' heights fall around the middle value but there are a few very short and a few very tall students.

Example

Sketch a graph to show your heart rate before, during and after a session of exercise

Label the axes with relevant information.

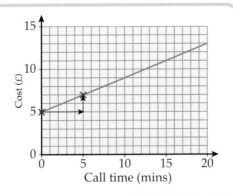

Steady rate

Increase in rate

Slow decrease back to original rate

Start   End of exercise   Time

● Straight-line graphs can be **interpreted** with more detail if you know the equation of the straight line.

The gradient represents the cost per minute.

Example

The graph shows the mobile phone charges used by one company. Explain their charging policy.

y-axis intercept is 5, gradient is $\dfrac{7-5}{5-0} = 0.4$
Equation is $y = 0.4x + 5$
Even if you don't make a call you are charged £5.
This could be line rental.
Every extra minute you spend on the phone increases your bill by £0.40
The company charges 40 pence per minute on calls.

Exercise 6h

1   Sketch a graph to illustrate each of these situations.
    In each case use the *x*-axis to represent time.
    a   The cost of petrol which is rising steadily.
    b   The temperature of a cup of tea over half an hour.
    c   The weekly amount of pocket money given throughout childhood.
    d   The speed of a parachutist as she jumps from a plane.
    e   The level of water in a bath as it is filled, used and emptied.

2   Sketch a graph to show the depth of water against time if water
    is flowing into each container at a constant rate.

a   **b**      **c** 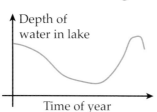   **d**

    e   Explain why the graph of volume against time would be the
          same for all the containers and sketch this graph.

3   Explain what is happening in each of the following sketch graphs.

a    **b**

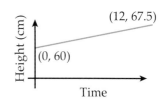

Problem solving

4   Nicholas hires a hall to hold his birthday party. The cost of the hall
    is found using the formula
        $y = 2x + 50$ x = number of people y = cost in pounds.
    a   Nicholas invites 50 friends but is not sure if they will all come.
          Using the formula given, plot a graph of number of people
          attending the party against the cost of hiring the hall.
    b   Use your graph to find the number of people if the cost is £78.
    c   Interpret the values of the gradient and the *y*-axis
          intercept on your graph.

5   The sketch graph shows the height of a baby during the
    first 12 weeks of its life.
    Find the equation of this line graph and interpret its meaning.

6   Investigate the meaning of IQ and use your findings to sketch a
    graph of IQ against time for a man ageing from 20 to 50 years.

6i Time series

○ A graph of repeated measurements of a quantity against time is called a **time series**.

Many time series show both regular short-term patterns as well as underlying long-term patterns.

Example

Describe the long term trend and any other patterns you see in this index data for retail food sales.

The food sales index gives the sales in each quarter as a percentage of the average sales in the whole of the year 2000.

p.138 >

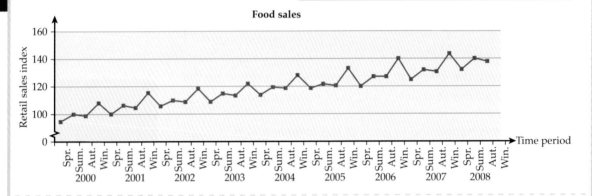

The long term trend is steadily upward (inflation).
Each year, there is a peak in the Winter, which includes Christmas, and a slight dip in the Spring.

○ A pattern which repeats each year is called a **seasonal variation**.
○ A longer term pattern is called a **trend**.

A trend describes how the 'average' data value changes.
Do not give detailed descriptions of fluctuations. Instead try to describe overall patterns in the data

Example

The graph shows three time series for the carbon dioxide emissions from industry, domestic use and transport since 1970. Describe the trends in the data.

Transport emissions have increased steadily, and doubled in the period shown.
Domestic use has varied quite a lot during the period but there is a slight general downward trend.
Industry has reduced its CO_2 emissions by about half over the 35 year period, though it has not been a steady rate of decrease.

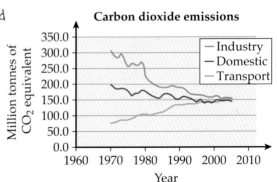

Carbon dioxide emissions

Exercise 6i

Problem solving

1 The average daily numbers of deaths or serious injuries in road traffic accidents, for each hour, on weekdays in 2002 is shown in the graph.

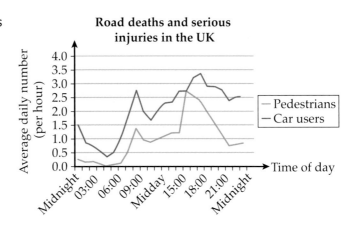

a What times of the day are there peaks in the number of accidents for both pedestrians and car users?

b Can you suggest why this is?

2 The graph shows the number of new cars (in thousands) registered each month during 1994–1996.

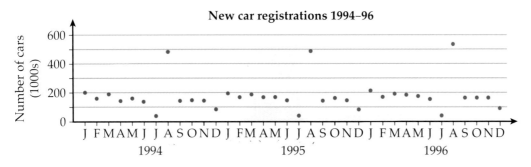

a Describe the long term trend for the data.

b There were about two million new cars registered each year in this period. Approximately what fraction of them were registered in August?

c After August, what was the second most popular month for new registrations?

> Before 1999 cars sold in August had new registration plates.

3 In 2001, the system for number plates was changed, so the time identifier changed twice every year. From the time series of new registrations during 2005–2007, which months do you think the changes occur? Why do you think the change was brought in?

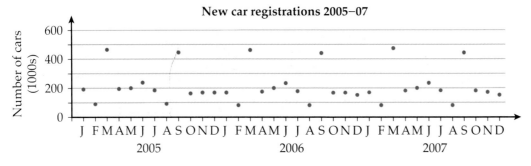

6j Exponential and reciprocal graphs

The equation $y = 2^x$ generates an **exponential sequence**.

1, 2, 4, 8, 16 and so on

⬤ Exponential graphs are of the form $y = a^x$, where a is a constant.

These patterns are often seen in the growth of bacteria in an infection.

This graph shows the growth of bacteria over 6 hours.

a   How many bacteria were present after 4 hours?

b   How many bacteria were present after 4.5 hours?

c   How long did it take the number of bacteria to reach 250?

a   160

b   230

c   about 4.7 hours

It can be tricky to read accurately off a graph.

Take care to be accurate when reading a graph.

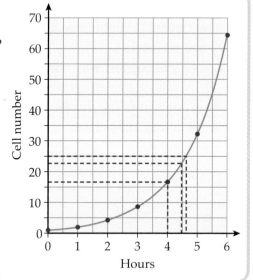

Not all the graphs you see are tidy straight lines or smooth curves.

Some graphs have gaps, for example they may have x-values which have no y-value. Other graphs can have more than one y-value for an x-value.

The graph shows the **reciprocal** function $y = \frac{1}{x}$.

a   Find y when x is 1.

b   Find y when x is 4.

c   Find x when y is −2.

d   What do you notice when $x = 0$?

a $y = 1$

b $y = \frac{1}{4}$ or 0.25   The reciprocal of 4 is $\frac{1}{4}$

c $x = -\frac{1}{2}$

d   There is no value!

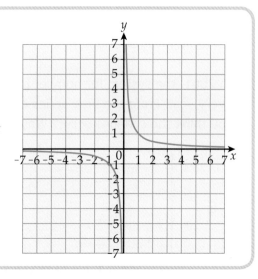

Exercise 6j

Problem solving

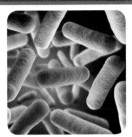
1 Bacteria reproduce by splitting in two. Beginning with one bacterium, this splits, after one hour, to make two bacteria and these split, after a further hour, to make four bacteria and so on.

  a Copy and complete the table to show the number of bacteria over a period of 10 hours.

Time	0	1	2	3	4	5	6	7	8	9	10
Bacteria	1	2	4								

  b Plot a graph to show the number of bacteria over time.

  c Use your graph to estimate how long it takes for there to be 1000 bacteria.

  d Can you suggest an equation for your graph?

  e How would the graph differ if a new strain of bacteria was discovered that splits into three to reproduce? Sketch your graph.

2 Copy and complete the following tables of values and use them to plot an exponential graph in each case.

Here are three possible variations of the power button.

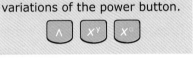

x	-4	-3	-2	-1	0	1	2	3	4
y									

  a $y = 3^x$ **b** $y = \left(\frac{1}{2}\right)^x$ **c** $y = 2^{-x}$

3 The population P, over time t years, of a herd of elephants is modelled using the formula $P = 4 \times 1.5^t$.

  a What is the initial population of the herd?

  b What is the population of the herd after

    i 1 year   **ii** 2 years?

  c Plot a graph of P against t for a 10-year period and use it to find the point in time at which the herd exceeded 100 elephants.

4 For each of the following functions

  i Copy and complete the table of values

x	-4	-3	-2	-1	0	1	2	3	4	5	6
y											

  ii Draw suitable axes and plot each graph, joining your points with a smooth curve in each case

  a $y = \frac{12}{x}$ **b** $f(x) = \frac{20}{x} + 1$ **c** $y = \frac{12}{x-2}$

5 It takes one person 10 days to build a conservatory. Plot a graph of how long it would take up to 10 people to build this same conservatory and find the equation of your graph.

Check out

You should now be able to ...

Test it ➡

Questions

✓ Plot graphs of linear functions and find gradients.	7	1, 2
✓ Find the equation of straight-line graphs.	7	3, 4
✓ Recognise and plot graphs of simple quadratic functions.	8	5
✓ Recognise and plot graphs of cubic functions.	8	6
✓ Plot and interpret distance-time graphs.	7	7
✓ Plot and interpret real-life and time series graphs.	7	8
✓ Read and interpret exponential and reciprocal graphs.	8	9

Language	Meaning	Example
Gradient	A mathematical measure of the steepness of a line.	The gradient of the line between (0, 0) and (4, 2) is 2
Line segment	A line between two set points.	The straight line between (0, 0) and (4, 2) is a line segment
Straight-line graph	A line joining a set of points set by a linear equation.	$y = 2x + 4$ is a straight-line graph
Quadratic function	A function where the highest power of x is 2.	$y = x^2 + 2x + 4$
Cubic function	A function where the highest power of x is 3.	$y = x^3 - x^2 + 2x + 4$
Time series	A graph showing how a measurement changes over time.	A patient's temperature chart is a time series
Exponential	A sequence that increases using a power of a number.	2, 4, 8, 16, 32...

1 Find the gradient of the straight line through the points (1, -2) and (-3, -4).

2 Find the gradient and the y-intercept of the straight lines with these equations.

   a $y = \dfrac{x}{4}$ **b** $y = 10 - 7x$

   c $y = -4(7 - 2x)$ **d** $8x + 2y = 10$

3 Find the equations of each of these straight lines.

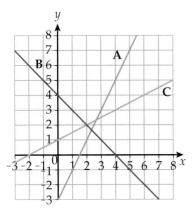

4 Give the equation of a graph that is parallel to $y = 3x - 1$.

5 **a** Plot the graph of $y = x^2 + 2x - 1$ for values of x from -4 to 2.

   b Write the equation of the line of symmetry of the curve.

   c State the coordinates of the minimum point on the curve.

6 Copy and complete the table of values for the cubic function $y = 4x - x^3$. Use your values to plot the graph of the function.

x	-3	-2	-1	0	1	2	3
4x							
-x³							
y							

7 A man walks away from his home at a speed of 1.5 m/s for 20 minutes, then turns around and runs home at a speed of 4 m/s. Draw a distance-time graph to show his journey.

8 Water is flowing into this container at a constant rate. Sketch a graph to show the depth of water against time.

9 **a** Draw the graph of $y = \dfrac{4}{x}$ and use your graph to approximate the value of x for which $y = 1.9$.

   b Draw the graph of $y = 3^x$ and use it to estimate the solution to $3^x = 10$.

What next?

Score		
	0 – 4	Your knowledge of this topic is still developing. To improve look at Formative test: 3C-6; MyMaths: 1312, 1153, 1314, 1316, 1322, 1184, 1070 and 1071
	5 – 7	You are gaining a secure knowledge of this topic. To improve look at InvisiPen: 263, 265, 271, 272, 273, 274, 275 and 278
	8 – 9	You have mastered this topic. Well done, you are ready to progress!

6a

1 Find the gradient of the line segment joining the following pairs of points.

    a (2, 6) and (4, 10)   **b** (7, 0) and (5, 8)

    c (-4, 5) and (-2, -1)   **d** (-1, -6) and (0, -5)

    e (0, 6) and (5, 11)   **f** (0, -3) and (-5, -9)

2 Find the value of a if the gradient joining the points $(-2, a)$ to $(a + 1, -10)$ is -2.

6b

3 **a** Write the equation $4x = 2y + 1$ in explicit form.

    b Copy and complete the table of values for the equation $4x = 2y + 1$

x	0	1	2
y			

    c Plot these values on a pair of coordinate axes and join them with a straight line.

    d Write the coordinates of the point where $4x = 2y + 1$ intercepts the y-axis.

    e Does the point (12, 24.5) lie on the line?

6c

4 Give the equation of a line which is

    a parallel to $y = 2x + 5$ **b** parallel to $3y = x - 1$

    c perpendicular to $y = 5 - 4x$ **d** perpendicular to $2x + 5y = 10$.

6d

5 **a** Copy and complete the table of values for the function $y = x^2 - 2x - 3$.

    b Plot the graph of $y = x^2 - 2x - 3$.

    c Write the equation of the line of symmetry of the curve.

    d Write the coordinates of the minimum point of the curve.

x	-2	-1	0	1	2	3	4
x^2	4			1			
$-2x$	4			-2			
-3	-3			-3			
y	5			-4			

6e

6 Plot a graph of $y = 2x^2 - x + 3$.

    Use your graph to find

    a the coordinates of the maximum or minimum point

    b the equation of the line of symmetry.

6f

7 Make a table of values and use it to plot the graph of $y = x^3 + x$.

    Use your graph to estimate the value of y when $x = 2.4$.

8 The graph shows the journey of a cyclist.

   a What was the speed of the cyclist in the first part of the journey?

   b For how long did the cyclist rest during the whole journey?

   c At what speed was the cyclist travelling at after $3\frac{1}{2}$ hours?

   d How far did the cyclist travel altogether?

   e Ignoring rest periods, what was the average speed of the cyclist for the entire journey?

   f True or false: After $1\frac{1}{2}$ hours the cyclist started to repeat part of the journey that he had just covered.

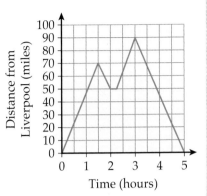

9 Make a sketch graph to show

   a the average daily number of calories consumed for the first 25 years of a person's life

   b temperature against time for 24 hours on a spring day in London.

10 The graph shows time series of the stocks of various types of fish in the North Sea since 1963. Describe the trends in the data.

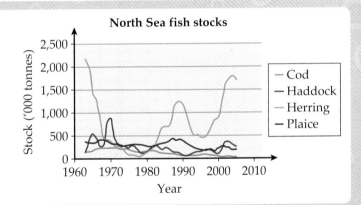

11 Match each graph with its equation.

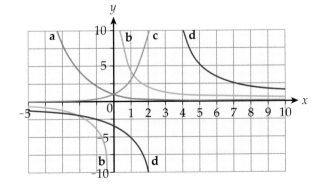

   i $y = \dfrac{4}{x}$

   ii $y = 3^x$

   iii $y = \dfrac{10}{x-3}$

   iv $y = 2^{-x}$

Katie and Jess are going to make bracelets and necklaces to sell on an online auction site.
They need to ensure that they keep their costs down, to help them run a profitable business.

Katie and Jess have found out the cost of materials from two suppliers.

Each supplier quotes prices for two types of bead—long ones or round ones.

They also quote prices for waxed cord or leather thread.

diameter x length
8mm x 16mm beads

diameter x length
12mm x 9mm beads

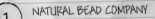

① NATURAL BEAD COMPANY
Postage and packing: £3.50 for any size of order

8 x 16mm beads	12 x 9mm beads
8p per bead	5p per bead
£1.30p per 20	80p per 20
£2.90 per 50	£1.80 per 50
£4.80 per 100	£3.00 per 100
£8.00 per 250	£5.00 per 250

Leather thread	Waxed cord
50p per metre	10p per metre
£11.95 per 50m	£4.95 per 100m

Task 1
a Katie and Jess want 2000 beads of each size.
Which supplier should they use? Show your workings out.
b They also want 50m of leather thread, as well as 50m of waxed cord. Again, decide which supplier they should use, showing clearly your workings out.

② BEAD-E-IZE
Free postage and packing. Minimum order charge of £10

8 x 16mm beads	12 x 9mm beads	Waxed cord
7p per bead	4p per bead	
£1.50p per 25	85p per 25	
£6.75 per 150	£3.75 per 150	
£19.00 per 500	£10.75 per 500	

Leather thread	Waxed cord	Leather thread
45p per metre	11p per metre	
£19.00 per 80m	£1.95 per 20m	
	£9.50 per 200m	

Task 2
Katie and Jess want their bracelets to be 16cm long, with an adjustable tie. Only ¾ of this length can be used for beads.

a How many long beads would fit on a bracelet?
b How many round beads would fit on a bracelet?
c If they just make bracelets, how many bracelets could they make?

Task 3
Katie and Jess want their necklaces to be 30cm long. Only ⅔ of this length can be used for beads.

a How many long beads would fit on a necklace?
b How many round beads would fit on a necklace?
c If they just made necklaces, how many necklaces could the girls make?
d (Harder) Research shows that they will sell twice as many bracelets as necklaces. How many of each would you recommend that the girls make?

Task 4

In an online auction, Katie and Jess will have to pay the website:
· for advertising their jewellery (listing fee)
· for selling their jewellery (selling fee)

The fees for selling the items in the online auction are:

Auction listing fee:	
starting price	fee
£0.01 – £0.99	0p
£1.00 – £4.99	15p
£5.00 – £14.99	25p
£15.00 – £29.99	50p

Auction selling fee:	
not sold	0p
sold	10% of selling price

a What are the benefits of starting an auction at 99p?
b What are the risks?

Task 5

If they want to, they can sell the items at a set price rather than an auction. The fees for this are:

Fixed price sales	
Listing fee	40p
Selling fee	8.5% of selling price

a If they sell bracelets for £4.99, are they better off starting an auction at £4.99 or using the fixed price option?
b What is the lowest selling price at which fixed price fees would be less than auction fees?

Task 6
Your challenge!

Decide what you would do if you were setting up an online jewellery business. Think about things such as:

· What designs you would make
· How many of each item you would make
  (Making more can save money on materials but lose more if the items don't sell)
· What the items would cost to make
· How much you would sell them for
· Whether you would sell as an auction or at a fixed price
  (don't forget to include postage and packing costs — you can look these up on popular websites like eBay)
· How much profit you would hope to make
· How much you would initially spend to get started
  (you might want to limit the amount so that you don't lose too much money if your items don't sell)

Present your decisions as a business plan, setting out all the details and the reasons for the decisions you make.

7 Decimal calculations

Introduction

Have you ever wondered how a computer manages to calculate so quickly? Computers are essentially a set of switches which can either be 'on' or 'off'. For that reason they read numbers in **binary code**. This uses just two symbols '1' and '0', so you can write the number 5 as 101. So, the computer 'sees' the number 5 as two switches turned 'on' (1) and one switch turned 'off' (0).

When the computer needs to calculate, it converts the numbers into binary and then uses patterns of logic gates (which can be either OR, NOT or AND) to add, multiply and perform other types of calculation.

What's the point?

Computers are very useful devices because they can perform lots of calculations very quickly. This enables them to do highly complex tasks in seconds that it could take humans millions of years to do!

Objectives

By the end of this chapter, you will have learned how to …
- Know and use the correct order of operations.
- Use a range of mental and written strategies for decimal calculations.
- Use a calculator for complex calculations.
- Interpret the calculator display.

Check in

1   Calculate these using an appropriate method.
   a   21 × 32.8   b   3.4 × 43.7   c   564 ÷ 24   d   321.3 ÷ 7
2   a   Use your calculator to find $\sqrt{20}$ and write your answer to 3 dp.
    b   Square your rounded answer, and explain why the result is not 20.

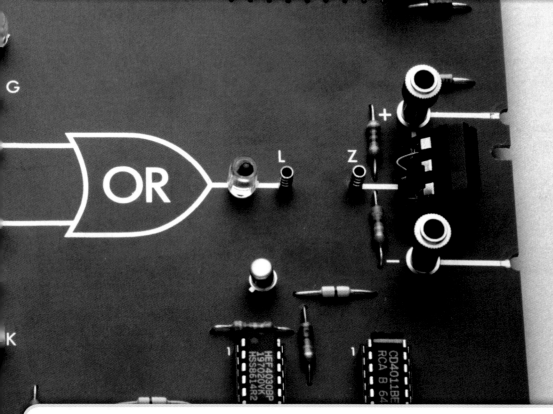

Starter problem

Here is a scientific calculator.
Unfortunately only the number 4 is working on the calculator.
However, all the operation and function keys are working.
Can you make all of the numbers from 1 to 20?

When you tackle complex calculations you need to use the correct order of operations.

Calculate: $\dfrac{(6.8 - 2.3)^2 - 5 \times 3}{7 \times 5 \div (2.5 - 0.5)^3 + 5}$

> Estimate first
> $(7 - 2)^2 - 15 = 25 - 15 = 10$
> $35 \div 8 + 5 \approx 36 \div 9 + 5 = 4 + 5 = 9$
> $10 \div 9 \approx 10 \div 10 = 1$

Work on the numerator and denominator separately using BIDMAS.

Numerator		Denominator	
$(6.8 - 2.3)^2 - 5 \times 3$		$7 \times 5 \div (2.5 - 0.5)^3 + 5$	
$= 4.5^2 - 5 \times 3$	B	$= 7 \times 5 \div 2^3 + 5$	B
$= 20.25 - 5 \times 3$	I	$= 7 \times 5 \div 8 + 5$	I
$= 20.25 - 15$	DM	$= 35 \div 8 + 5$	DM
$= 5.25$	AS	$= 4.375 + 5$	DM
		$= 9.375$	AS

Combining

$\dfrac{5.25}{9.375} = \dfrac{21}{4} \div \dfrac{75}{8}$

$= \dfrac{\overset{7}{\cancel{21}}}{\underset{1}{\cancel{4}}} \times \dfrac{\overset{2}{\cancel{8}}}{\underset{25}{\cancel{75}}}$

$= \dfrac{14}{25} = 0.56$

You could perform part or all of the calculation in the example on a calculator.

▶ Calculate the denominator (use the brackets and x^y keys). | `9.375` |
▶ Store your answer in the memory (often Min or STO). | `9.375 MIN` |
▶ Calculate the numerator (use the brackets and x^y keys). | `5.25` |
▶ Divide the answer by the recalled memory (often MR or RCL). | `0.56` |

Knowing inverse operations can help you to solve problems.

If $x^3 = 140$, find the value of x.

Read the equation.　　　$x^3 = 140 \Rightarrow$ A number cubed equals 140.

Reverse the operations.　$x = \sqrt[3]{140}$.

Use a calculator.　　　　$= 5.19$ to 2 decimal places.

> The cube root key often looks like .

Exercise 7a

1 Work out these expressions.

  a $3 \times 4 + 6 \times 2$ **b** $(4 + 2)^2 - 3$

  c $6 \times 2 \div (5 - 3)^2$ **d** $3.5^2 - 4 \div 2$

2 Add one pair of brackets to each of the following to make the equations correct.

  a $6 + 7^2 - 9 - 4 = 50$

  b $7 \times 6 \div \text{-}4 - 10 = \text{-}3$

  c $8 \times 7 + 8 - 7 = 64$

  d $\text{-}9 \times 3 - 5 \times 2 = 36$

3 Insert the operations that produce the answer shown in each calculation below.

  a $(8 \, ? \, 7) \, ? \, 6 \, ? \, 5 = 18$

  b $1 \, ? \, 2 \, ? \, 3 \, ? \, 4 = \text{-}1$

  c $(10 \, ? \, 9 \, ? \, 8) \, ? \, 7 = \text{-}49$

  d $(5 \, ? \, 10 \, ? \, 15) \, ? \, 2 = 310$

4 Work out each of these when $m = 6.5$, giving your answer to 2 dp.

  a $\dfrac{4m^2 - 6(m - 5.7)}{2m}$

4 b $\sqrt{(m^2 - m)} - 3(4 - \sqrt{m})$

  c $\dfrac{(\text{-}m)^3 - 5(m - 5.4)}{7 - m}$

5 The expression

$6(7 - 8) = 6(\text{-}1) = \text{-}6$

can be thought of as

$6(7 - 8) = 6 \times 7 - 6 \times 8 = \text{-}6$

Complete the expressions below.

  a $10(7 - 3) = 10 \times ? - 10 \times ? = 40$

  b $8(? + ?) = 8 \times 7 + ? \times 4 = 88$

  c $7(? - ?) = ? \times 10 - ? \times ? = 28$

  d $6(a + b) = 6a + ?$

  e $5(f + \text{-}5) = ?$

6 $\dfrac{(6 + \text{-}7)^2 - 6 \times \text{-}9}{7^2 - 8}$

  a Estimate the answer.

  b Work out the exact answer to the expression, giving your answer to 2 dp.

Problem solving

7 The key sequence for calculating $\sqrt[3]{1728}$ is $\boxed{\sqrt[3]{\,}}\,\boxed{1}\,\boxed{7}\,\boxed{2}\,\boxed{8}\,\boxed{=}$

However, there is another possible key sequence using the y^x key. Can you discover it? Write it out using the notation shown above.

8 a Find an integer between 70 and 80 that is made up by using the following -3, 9, -5, ( ), ^{2}

  b Find an integer between 1 and 10 that uses the number 5 four times and the operations $+$ and $\div$ once each.

9 Use the digits 1 to 5, each of the operations $\times$, $\div$, $+$, $-$ once only, and two pairs of brackets to produce the answer $\frac{1}{15}$.

10 a Explain in words the difference between the expressions: $(\text{-}h)^2$ and $\text{-}h^2$

  b Show using values for h that these two expressions produce different results.

  c Are there any values for which $(\text{-}h)^2$ and $\text{-}h^2$ are equal? Explain your answer.

  d Is there a difference between $(\text{-}h)^3$ and $\text{-}h^3$?

MyMaths.co.uk   Q 1167   SEARCH

> You can use a standard column procedure for adding and subtracting numbers with differing numbers of decimal places.

Calculate 7239 + 470.004 + 0.0032

Add zeros so that all the numbers have the same number of decimal places.

```
  7239.0000
   470.0040     Set out the calculation in columns
     0.0032     Align the decimal points
  7709.0072
```

> There are standard methods for efficiently multiplying and dividing decimals.

Use a standard method to calculate

a   0.0632 × 0.047

b   0.0735 ÷ 0.68 to 3 sf

Estimate the answer.

a $0.0632 \times 0.047 \approx 0.06 \times 0.05 \approx 0.003$

b $0.0735 \div 0.68 \approx 0.07 \div 0.7 \approx 0.1$

Change the decimal calculation into an **equivalent integer calculation**.

0.0632×0.047

$= (632 \div 10\,000) \times (47 \div 1000)$

$= (632 \times 47) \times 10^{-7}$

$\dfrac{0.0735}{0.68} = \dfrac{735}{10\,000} \times \dfrac{100}{68}$

$= (735 \div 68) \div 100$

```
     632
   ×  47
   25280      40 × 632
    4424       7 × 632
   29704
```

```
         10.80...        Write 0 as 68
   68) 735.00            is too big to divide.
      -68                    1 × 68
       55.0
      -54.4                  .8 × 68
        .60
```

Check answer against estimate.

$0.0632 \times 0.047 = 29704 \times 10^{-7}$

$= 2.9704 \times 10^{-3}$

$0.0029704 \approx 0.003$ ✓

$.0735 \div 0.68 = 10.80... \div 100$

$= 0.108 \ (3\text{ sf})$

$0.108... \approx 0.1$ ✓

Exercise 7b

1 Calculate these using an appropriate method.

a 5.63 + 435.2 + 6

b 64.3 + 216.8 + 50.32

c 7.12 + 0.4 + 7 + 0.46

d 425.1 + 38.16 + 0.027

e 243.5 − 6 − 37.8

f 57.6 + 284.8 − 61.92

g 47.3 − 2.9 − 14

h 354.68 − 88 − 4.7

2 Calculate these using a written method.

a 32 × 4.37   b 27 × 1.64

c 63 × 35.7   d 47 × 0.24

e 75 × 0.96   f 32 × 0.75

g 7.1 × 36.1   h 4.9 × 52.7

i 9.3 × 4.48   j 0.83 × 4.45

k 0.63 × 4.71   l 0.18 × 88.4

m 0.32 × 0.0311   n 0.28 × 0.678

o 0.63 × 0.504   p 0.045 × 0.63

q 0.063 × 0.095   r 0.084 × 0.00572

s 0.00725 × 0.078 t 0.0238 × 0.00056

3 Calculate these using an appropriate method.

Give your answers to three significant figures.

a 384 ÷ 5.9   b 275 ÷ 1.9

c 835 ÷ 6.4   d 513.4 ÷ 0.4

e 713.2 ÷ 0.7   f 419.3 ÷ 0.6

g 326.7 ÷ 0.03   h 586.2 ÷ 0.08

i 395.4 ÷ 0.09   j 0.036 ÷ 0.0077

k 0.017 ÷ 0.0023 l 0.083 ÷ 0.00091

m 0.084 ÷ 0.64   n 0.783 ÷ 6.5

o 0.0624 ÷ 0.56   p 0.375 ÷ 0.49

q 0.073 ÷ 0.00392 r 0.0529 ÷ 0.36

Problem solving

4 a Hanif buys 0.064 tonnes of coal which costs £103.45 per tonne.
   How much does the coal cost?

  b Jason measures a microchip. It is 0.0375 m long and 0.0013 m wide.
   What is the area of the microchip in square metres?

5 Give your answers to these calculations as decimals to 3 sf.

  a Barry the plankton is 0.037 cm long. His friend Gary is
   0.027 cm long. How many times longer is Barry?

  b A section of a DNA molecule measures 34 angstroms
   long by 21 angstroms wide.
   What is the ratio of length to width for this section
   of DNA molecule?
   (1 angstrom = 1.0×10^{-10} metres)

21 angstroms

34 angstroms

6 You can find the square root of 5 using the sequence generated
   by this term-to-term rule.

  a Start with the estimate $x_1 = 2$ and work out x_2.

$$x_{n+1} = \frac{\left(x_n + \frac{5}{x_n}\right)}{2}$$

  b Generate further terms in the sequence $x_1, x_2, x_3, x_4, \ldots$ until you can give $\sqrt{5}$ to 3 sf.

  c Find the square root of other numbers using this method.

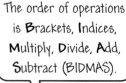
The order of operations is Brackets, Indices, Multiply, Divide, Add, Subtract (BIDMAS).

● When a calculation contains more than one operation, you must do the operations in the correct order.

Example

Calculate $\dfrac{(7.6 \times 10^{15}) - (3.2 \times 10^7)^2}{3.4 \times 10^4 + \sqrt{1.44 \times 10^8}}$

Work out any powers or roots

$\dfrac{(7.6 \times 10^{15}) - (1.024 \times 10^{15})}{3.4 \times 10^4 + 1.2 \times 10^4}$

$= \dfrac{6.576 \times 10^{15}}{4.6 \times 10^4}$

$= 1.429565217 \times 10^{11} = 1.4 \times 10^{11}$ (2 sf)

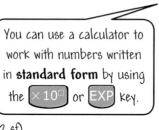
You can use a calculator to work with numbers written in **standard form** by using the $\boxed{\times 10^{\square}}$ or $\boxed{\text{EXP}}$ key.

● Always set out a problem as a set of mathematical calculations. Then use the functions on your calculator to work out the answer.

Example

This diagram shows a circle inside a square.
The circle touches the edge of the square.
What percentage of the diagram is shaded?

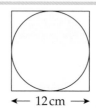

← 12 cm →

Area of circle = π × radius² = π × 6² = 113.0973355 cm²
Area of square = length² = 12² = 144 cm²

Percentage shaded = $\dfrac{\text{area of circle}}{\text{area of square}} \times 100\%$

$= \dfrac{113.0973355}{144} \times 100\%$

Percentage shaded = 0.7853981634 × 100% = 78.5% (3 sf)

Leave all of the digits of your calculation to avoid rounding errors.

Example

Find the value of $12^{\frac{2}{3}}$

Enter $12^{\frac{2}{3}}$ on your calculator.

$12^{\frac{2}{3}} = 5.24$ (3 sf)

You could also use the $\boxed{\text{ANS}}$ key.

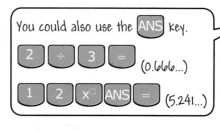

Exercise 7c

1 a Use your calculator to evaluate

   i $16^{\frac{1}{2}}$ **ii** $8^{\frac{2}{3}}$ **iii** $16^{\frac{3}{4}}$

   iv $9^{\frac{3}{2}}$ **v** $27^{\frac{2}{3}}$ **vi** $81^{-\frac{1}{4}}$.

  b What does the denominator of a fractional power do?

  c What does the numerator of a fractional power do?

2 Evaluate

  a $(3.14 \times 10^3)^2$ **b** $\sqrt{6.89 \times 10^5}$

3 Calculate the following, giving your answers to 2dp.

  a $\dfrac{(5.1 \times 10^2 + 3.2 \times 10^3)^2}{(2.1 \times 10^2)^3}$

  b $\dfrac{(3.2 \times 10^4 - 1)\times(7.05 \times 10^{-2})^2}{9 - 3.3 \times 10^{-2}}$

  c $\dfrac{4 \times 10^7 \times (1.8^{\frac{3}{2}} + 4.2^{\frac{1}{2}})^{\frac{2}{3}}}{1.05 \times 10^{-5}}$

  d $\dfrac{(3.03 \times 10^{26})^2 \times 3.14}{2}$

Problem solving

4 Rachel says that when a circle fits just inside a square it will always cover about $\frac{3}{4}$ of the square.

Work out the percentage of the square covered by the circle for squares of different lengths.

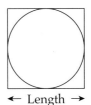

← Length →

5 Ryan says that the square of a number is always bigger than the number. For example, the square of 3 is 9.

  a Find some numbers for which this statement is not true.

  b Describe the type of number you found in part **a**.

6 a The speed of the fastest spacecraft is 4×10^4 km per hour. How long would it take for this spacecraft to travel to

   i the Moon (distance = 238855 miles)

   ii Proxima Centauri (distance = 4.0×10^{13} km)?

  b How fast would the spacecraft have to travel to reach each destination in

   i 1 year   **ii** 1 week?

Did you know?

The fastest speed possible is, c, the speed of light. This is used in Einstein's famous equation $E = mc^2$.

7 Karen wants to investigate the graphs of these different functions.

$y = \dfrac{1}{x}$ $y = \dfrac{1}{(x - 4)}$ $y = \dfrac{1}{x^2}$

x	0.5	1	1.5	2	2.5	3
y						

She completes a table of values for each function.

  a Copy and complete a table for each function.

  b Draw a graph for each function.

  c Use some more values for x which are either less than 0.5 or greater than 3 and plot these points on your graphs.

   Describe what is happening to your graph

   "As x gets smaller ..."   "As x gets bigger ..."

  d Try some negative values for x. Describe what happens.

MyMaths.co.uk   Q 1167   **SEARCH**

When you use a calculator for a division you may have a **remainder**.
It can be written as a whole number, a fraction or a decimal.

At the local supermarket giant baked beans
are on special offer.
You can buy a six-pack of tins for £7.89.
Xin buys a six-pack.
How much did she pay for each tin of beans?

Estimate $8 ÷ 6 ≈ 8 ÷ 4 ≈ 2$
Using a calculator, $7.89 ÷ 6 = 1.315$
Each tin cost £1.32

> Decide how to interpret the answer in the context of the problem. As the answer is money, round to 2 dp

● You can convert decimal remainders into whole numbers
by multiplying by the divisor.

Convert 5000 seconds into hours, minutes and seconds.

Estimate $5000 ÷ 60 ≈ 5000 ÷ 50 ≈ 100$
Using a calculator, convert the seconds into minutes.
$$5000 ÷ 60 = 83.333...$$
Multiply the remainder by the divisor.
$$0.333... \text{ minutes} = 0.333... × 60 \text{ seconds}$$
$$= 20 \text{ seconds}$$
$$5000 \text{ seconds} = 83 \text{ minutes and } 20 \text{ seconds}$$
$$= 1 \text{ hour, } 23 \text{ minutes and } 20 \text{ seconds}$$

> Divide by 60 to change seconds to minutes.

> Change the remainder into a whole number by multiplying by 60.

Ciara and Celine convert $\frac{6}{7}$ into a percentage, giving
their answer to 1 dp.
Ciara $6 ÷ 7 = 0.85714...$ Celine $6 ÷ 7 = 0.85714...$
$= 0.9$ $= 85.714...\%$
$= 90.0\%$ $= 85.7\%$
Explain why their answers are different.

Ciara has rounded her answer to 1 decimal place *during* the calculation.
Celine has rounded *at the end of* the calculation.
Celine's answer is more accurate.

> Round the answer at the end of the calculation to avoid making rounding errors.

Exercise 7d

1 Without using a calculator choose the most likely answer
for each question by doing a mental estimate.
  a $(38)^2 =$ 54872   or   1444   or   76
  b $130 \div 0.45 = 58.5$ or   244.2   or   288.9
In each case explain the reasoning behind your choice.

2 Do these divisions using your calculator.
Give the answer in the form stated.
  a £30 ÷ 7   (a decimal to 2 dp)
  b 30 cakes ÷ 7   (with a whole number remainder)
  c 30 pies ÷ 7   (a fraction)

3 Convert these measurements to the units in brackets.
  a 2458 cm   (into m and cm)
  b 3876 seconds   (into minutes and seconds)
  c 43155 mm   (into m, cm and mm)

Problem solving

4 Solve these problems.
Give your answer in a form appropriate to the question.
  a Dale sells free range eggs.
  She has 155 eggs. She packs them into boxes of 12.
  How many boxes of eggs does she need to pack the eggs?
  b At a charity event, 12 competitors eat a total of 155 pizzas.
  Each competitor eats exactly the same amount of pizza.
  How much pizza did each person eat?

5 Solve these problems.
Give your answer in a form appropriate to the question.
  a Igor buys 7 kg of sugar.
  He wants to put the sugar into 1.5 kg jars.
  How many jars will he need?
  How much sugar will there be in the last jar?
  b The Year 9 students at Oxford Sports College are going on a
  trip to a theme park.
  There are 189 students and 16 staff going on the trip.
  Each coach can hold 45 people.
  How many coaches should be ordered?

6 Convert 99999999 seconds into years, weeks, days, hours, minutes and seconds.

Check out

You should now be able to ...

Test it ➡

Questions

✓ Know and use the correct order of operations.	7	1
✓ Use a range of mental and written strategies for decimal calculations.	6	2 – 5
✓ Use a calculator for complex calculations.	8	6
✓ Interpret the calculator display.	7	7, 8

Language	Meaning	Example
Equivalent integer calculation	Using powers of 10 to simplify a calculation.	$\dfrac{0.004}{0.0002} = \dfrac{40}{2} = 20$
Order of operations	A standard order of doing operations in a calculation: **B** brackets **I** powers/indices **D** division **M** multiplication **A** addition **S** subtraction.	$3 + (4^2 - 7 \times 2)$ $= 3 + (16 - 14)$ $= 5$
Rounding errors	Errors that occur by rounding an intermediate answer in a calculation.	$2 - 3 \div 2 = 2 - 1.5 = 0.5$ ✓ With intermediate rounding $= 2 - 2 = 0$ ✗

1 Work out each of these using a calculator. Give your answers to 2 dp.

   a $(5.3)^3 - 2.7 \div 3.14$

   b $(6.9 - 4.02)^2 \div 2$

   c $4\frac{1}{8} \times (6.3 - 2.5)$

   d $\dfrac{(8.3 - 3.1)^3}{8}$

   e $9 + \dfrac{3}{4} \div 3.6$

   f $\dfrac{2.1}{5.6^3} - 9.4^2$

2 Work out these without using a calculator.

   a $0.383 + 5.48 + 89.1$

   b $103.2 + 25.11 + 1.57$

   c $9.56 - 0.892 + 0.09$

   d $44.42 - 0.147 - 0.0319$

3 Calculate these using a written method

   a 33×5.52

   b 71×0.49

   c 2.9×1.56

   d 0.47×0.0672

4 Use a written method to work out these products.

   a 23.1×5.2

   b 0.52×7.9

   c 6.96×3.2

   d 1.34×5.39

5 Calculate these, give your answer to 3 sf.

   a $755 \div 5.7$

   b $633.6 \div 0.92$

   c $486.3 \div 0.07$

   d $0.019 \div 0.71$

   e $0.037 \div 0.00741$

6 Calculate the following, give your answers to 2 dp.

   a $\dfrac{(1.9 \times 10^3 + 7.8 \times 10^2)^3}{(2.6 \times 10^{-3})^{-4}}$

   b $\dfrac{(8.1 \times 10^3 - 6) \times (9.03 \times 10^{-2})^3}{8 - 2.7 \times 10^{-3}}$

7 The nearest the planet Mars has ever been to earth is 5.6×10^7 km and the furthest is 4.01×10^8 km.

   a What is the difference between these two distances?

   b What is the maximum amount of time it could take to travel to Mars if your average speed is 3.8×10^4 km/h? Give your answer in days and hours and minutes

8 Do these divisions on a calculator and give your answer in the form stated.

   a £50 ÷ 7 (£ and p)

   b 30 cm ÷ 9 (nearest mm)

   c 12 days ÷ 15 (hours and minutes)

What next?

Score			
	0 – 3		Your knowledge of this topic is still developing. To improve look at Formative test: 3C-7; MyMaths 1007, 1011 and 1167
	4 – 6		You are gaining a secure knowledge of this topic. To improve look at InvisiPen: 132, 133 and 134
	7 – 8		You have mastered this topic. Well done, you are ready to progress!

MyMaths.co.uk

1 Use BIDMAS to evaluate each of these expressions.

   a $16 + 5 \times 4 - 42 \div 7$

   b $72 \div 9 - 15 + 7.4 \times 3$

   c $54 \div 6 - 15 + 3 \times (17 - 12)^2$

   d $16 - \sqrt{45} \div 9 + (13 - 5) \times 3 - 32$

   e $12 \times (25 - 13) \div 48 + 15$

   f $120 \div (15 - 7)^2 + 9 \times 3 - \sqrt{35}$

2 Evaluate each of these expressions using BIDMAS.

   a $\left(\dfrac{1}{2} + \dfrac{3}{4}\right) \times 12 - 5 \times \dfrac{1}{3}$

   b $\dfrac{1}{2} + \dfrac{3}{5} \times 15 - \dfrac{3}{4}$

   c $\dfrac{5}{8} \div 5 + 4 \times \dfrac{3}{4} - \dfrac{1}{8}$

   d $\left(\dfrac{1}{2}\right)^2 + 5 \times \dfrac{1}{4} - \dfrac{1}{2}$

3 Write a calculation that uses BIDMAS and gives an answer of 56.

4 Use BIDMAS to evaluate each of these expressions.

   a $\sqrt{56} \div 8 - 4 + 4 \times \sqrt{\dfrac{1}{4}}$

   b $24 - 3 \times 5 + \sqrt{108} \div (12 - 9)^2$

   c $\sqrt{(19 + 24)} + 5 - \sqrt{\dfrac{1}{9}} \times 18$

   d $(35 + 19) \div (13 - 11.5)^2 \times \sqrt{55}$

   e $(19 + 36) \div \sqrt{(8 - 3)} \div (8 - 5)^2$

   f $\sqrt{(6 \times 5 + 170)} \div 50 - (4 - 1)^3$

5 Evaluate each of these expressions using BIDMAS.

   a $\dfrac{3}{4} \div 3 + \dfrac{3}{8} \times 16 - \dfrac{1}{2}$

   b $1.5 \div 3 + \dfrac{1}{4} \times 7 - \dfrac{3}{4}$

   c $\dfrac{1}{2} + \left(\dfrac{1}{2}\right)^3 \times 8 + \dfrac{3}{2} \div 3$

   d $\left(\dfrac{2}{3}\right)^2 \times 18 + \dfrac{3}{4} \div 3 + 5$

6 Write a BIDMAS calculation, with brackets that gives an answer of 148.

7 Calculate.

a 27.6 ÷ 0.4	**b** 38.5 ÷ 0.5	**c** 22.5 ÷ 0.3	**d** 46.9 ÷ 0.7
e 138.6 ÷ 0.6	**f** 569.6 ÷ 0.8	**g** 463.5 ÷ 0.5	**h** 304.2 ÷ 0.3
i 366.1 ÷ 0.7	**j** 5073.5 ÷ 0.5	**k** 283.4 ÷ 0.2	**l** 146.7 ÷ 0.9
m 5062.6 ÷ 0.2	**n** 357.6 ÷ 0.8	**o** 4226.5 ÷ 0.3	**p** 1337.6 ÷ 0.4

8 Calculate.

a 38.6 × 0.8	**b** 132.4 × 0.7	**c** 204.8 × 0.3	**d** 166.5 × 0.4
e 304.6 × 1.2	**f** 223.6 × 1.4	**g** 15.8 × 6.2	**h** 131.4 × 4.5
i 24.6 × 3.8	**j** 18.4 × 3.7	**k** 37.8 × 5.4	**l** 65.2 × 7.2

9 Calculate.

a 45.6 ÷ 0.4	**b** 157.8 ÷ 0.6	**c** 487.5 ÷ 0.5	**d** 743.4 ÷ 0.7
e 46.2 ÷ 1.2	**f** 400.8 ÷ 1.5	**g** 65.7 ÷ 1.8	**h** 140.64 ÷ 2.4
i 354.2 ÷ 1.4	**j** 471.25 ÷ 2.5	**k** 90.24 ÷ 1.6	**l** 846.12 ÷ 2.2

10 A calculation was answered in this way

333.84 ÷ 2.6 = 100.2

Show how a check will tell you that the answer is wrong.

11 a Erik buys 0.43 tonnes of steel. The steel costs £186.23 per tonne.
How much does the steel cost?

b Lilian designs microcircuits. A particular chip is 0.0405 m long and 0.0024 m
wide. What is the area of the microchip in square metres?

12 Calculate these. Give your answers to 3 sf. where appropriate.

a $\dfrac{3.4^2 - 2.9}{4.1 + 0.23}$
  b $\sqrt{\dfrac{7.8^3 - 52}{5.5^2}}$
  c $2\pi\sqrt{\dfrac{0.9}{4.1^{\frac{1}{2}}}}$

13 Use your calculator to solve these problems.

a Sezer has 110.5 hours of music downloaded on her computer.
She accidentally deletes 80.15 hours of her music (luckily she has a back-up!)
How much music does Sezer have remaining on her computer?
Give your answer
  i in hours and minutes
  ii in minutes.

b Véronique puts carpet in her bedroom.
The bedroom is rectangular with length 4.23 m and width 3.6 m.
The carpet cost £6.79 per m² and can only be bought by the square metre.
  i Calculate the floor area of the bedroom.
  ii Calculate the cost of the carpet that is needed to cover the floor.

8 Statistics

Introduction

Watching TV is as bad for you as smoking cigarettes—every 30 minutes of watching TV shaves 11 minutes off the rest of your life!

This frightening headline certainly makes you think twice about watching TV, but statistics are not always what they seem. Politicians, advertisers and businesses often use techniques like this to represent statistics in order to support the ideas they want you to believe. Watching TV does not really shave 11 minutes off the rest of your life, but people who watch TV for most of their waking day tend to be less active and it is this lack of exercise which is unhealthy.

What's the point?

An appreciation of statistics enables you to analyse the sheer wealth of advertising and messaging that you receive every day, and work out what it really all means.

Objectives

By the end of this chapter, you will have learned how to ...

- ● Design experiments and surveys to test a hypothesis and collect data.
- ● Draw a frequency polygon.
- ● Find trends using moving averages.
- ● Estimate the mean from a grouped frequency table.
- ● Interpret a scatter diagram.
- ● Draw and use a cumulative frequency graph.
- ● Compare distributions.
- ● Use box plots to make comparisons between data sets.

Check in

1   The table shows the number of bullfinches seen at noon in a garden during one month.

Number of bullfinches	0	1	2	3	4	5	6	7
Frequency	8	3	7	4	4	0	3	1

For the number of finches calculate   **a** the mode   **b** the median
                                       c the range **d** the mean.

2   The table shows the ages of members of the teaching staff in an outdoor activity centre.

Age, x years	$20 \leq x < 25$	$25 \leq x < 30$	$30 \leq x < 35$	$35 \leq x < 40$	$40 \leq x < 45$
Frequency	6	8	5	2	1

a   Draw a bar chart for the data.

b   What is the modal class for the ages of the staff?

Starter problem

Tall people can jump further.

Investigate if this is true.

8a Planning a statistical survey

People can ask many questions but how often do they look at evidence?

If there is evidence you can use statistics. You must refine your question into a **hypothesis** that can be tested using data.

Then you must obtain data to test your hypothesis using a **survey** or an **experiment**.

▲ Who is the best band ever?

- **Primary data** is data you collect yourself.
- **Secondary data** is data someone else has collected.

▲ Are genetically modified crops safe?

Example

Do boys or girls do better in GCSE Maths exams?

a What do you think? Write a hypothesis.

b Identify **i** primary data and

ii secondary data you could use to investigate your hypothesis.

For each hypothesis it is possible to find evidence to test it against.

a Possible hypotheses are

'Boys do better than girls in GCSE Maths exams'

'There is no difference between boys and girls in GCSE Maths exams'

b **i** You could survey people who have taken the exams at your school.

ii Your school may provide data for previous years' exam results or you could use published national data.

Data to investigate statistical questions can be collected by a **survey** or an **experiment**.

Surveys include using questionnaires to seek opinions and taking measurements, for example, to investigate whether boys or girls are taller at age 14.

In an experiment one of the variables (the **explanatory** or **control variable**) will be controlled while its effect on the **response variable** is observed.

Example

Do people's reaction times increase after exercise?

a Should you use an experiment or a survey to investigate this question?

b Describe how you would carry out the investigation.

Exercise is the explanatory variable and the reaction time is the response variable.

a An experiment.

b Take a number of different people and record their normal reaction times and their reaction times after different amounts of exercise.

Exercise 8a

Problem solving

1 Write a hypothesis for each of the following questions.

   a Are boys or girls absent from school more often?

   b Do students have more homework in Year 11 than in Year 10?

   c Do people who live near nuclear power stations suffer more from cancer?

   d Will there be a change of government at the next general election?

   e Do boys or girls send and receive more text messages?

   f Do more students travel to school by car now than last year?

   g Do teenagers watch more television now than 10 years ago?

   h Do more students go to university now than 5 years ago?

You should always take time to plan data collection carefully. It is usually much harder to obtain information that was initially missed out.

2 For each hypothesis in question 1 identify sources of

   i primary

   ii secondary

   data that you could use.

3 For each of the following

   i explain whether you should investigate it using an experiment or a survey

   ii describe how you would do the experiment or decide who to survey.

   a Does practice improve people's performance on Sudoku puzzles?

   b How often do people go to the cinema?

   c Are you more likely to see at least one 6 when you throw two fair dice or at least 6 heads when you toss 10 fair coins?

   d What is the most popular holiday destination?

   e Do a majority of teenagers download music from the internet illegally?

   f Do people react more quickly with their left or with their right hand?

Did you know?

Sudoku was invented in 1979 by the American architect Howard Garns and by 2005 had swept the world. There are 5.472 billion possible Sudoku grids.

4 News agencies often commission surveys to try to measure the popularity of political parties.
What methods do polling agencies use and how reliable do you think their results are?

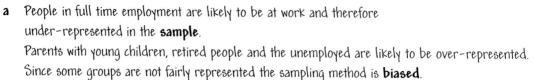

A survey is taken to find out the proportions of people unemployed, in part-time work or in full-time work.

a Explain why asking people in a town centre at 11 am on Tuesday morning is not a good idea.

b How could you improve the survey?

---

a People in full time employment are likely to be at work and therefore under-represented in the **sample**.

Parents with young children, retired people and the unemployed are likely to be over-represented.

Since some groups are not fairly represented the sampling method is **biased**.

b Doing the survey at the weekend when more people who have a job are around would be better.

Also carry out the survey in more than one location.

⬤ If the proportions of types of people in a sample are not the same as those in the whole **population** then the sample is biased.

Possible cause of bias include

▶ Only surveying people in one place at one time.

▶ Not matching the sample to the population.
For example, a survey about the reliability of mobile phones will be biased if you only include phones less than six months old.

▶ Asking sensitive questions without ensuring confidentiality.
For example, 'Have you ever taken drugs?'

▶ Poor response rates to surveys.
You don't know if the people who don't respond are similar to those who do respond.

▶ Asking leading questions.
For example, 'Do you agree that all children should learn to swim as soon as possible?'

Radio and online surveys are distorted because of the self-selecting bias of the participants.

Exercise 8b

Problem solving

1 A school wants to know how many books students read. The Deputy Head asks the librarian to find out, so she surveys the next 40 students who visit the library.

   a Explain why this will give a biased sample and whether it is likely to over- or under-estimate the number of books students read.

   b Suggest a better way for the librarian to take her sample.

2 Carole wanted to investigate whether wealthy people holiday abroad more often than less well-off people.
   She sent out a survey to 100 people but did not supply an envelope or stamp.

   a Give two reasons why this sample is likely to be biased.

   b Suggest how Carole could improve the survey.

Carole's Holidays Survey

Name: _____

Age: _____

Salary: _____

In the last two years how often have you been abroad on holiday?

_____

Please post your reply to the address supplied.

3 A town wants to know how far people in the town travel to work. They asked Keith to investigate.
   The next day, he asked all the people on the bus that he takes to work.

   a Explain why this will give a biased sample and whether it is likely to over or under estimate the distance people travel.

   b Suggest a better way for Keith to take his sample.

4 A council wants to do a survey on gender roles in local households. It decides to select a property at random and to take the first named person on the council tax register at that property.

   a Explain why this is not a good way to take the sample.

   b Could you give any way to improve the survey?

5 Most households get junk mail which may include surveys sent out by companies. If you can get hold of some of these, decide if any of them have any sources of possible bias.
   Do the companies offer any incentives (such as entry into a prize draw) to encourage people to take part?

8c Frequency diagrams

Histograms and **frequency polygons** are types of frequency diagrams. A histogram is a bit like a bar chart, but with no gaps between the bars.

Frequency polygons make it easy to see the shape of a distribution.

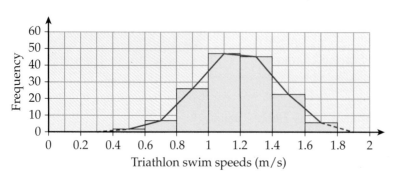

Triathlon swim speeds (m/s)

> A frequency polygon joins up the midpoints of the tops of the bars.

● The **modal class** is the interval with the highest frequency.

> The modal class is 1.0 to 1.2 m/s.
> The estimated range is 1.8 – 0.4 = 1.4 m/s.

● The **estimated range** is the largest possible data value minus the lowest possible data value.

You can draw a frequency polygon without drawing a histogram.

Example

Draw a frequency polygon to illustrate the distances (in metres) which a young golfer hit while trying some practice shots with a wedge.

Distance, d	$60 \leqslant d < 65$	$65 \leqslant d < 70$	$70 \leqslant d < 75$	$75 \leqslant d < 80$	$80 \leqslant d < 85$
Frequency	6	14	35	22	8

For the main frequency polygon, plot (62.5, 6), (67.5, 14), (72.5, 35), (77.5, 22) and (82.5, 8). To complete the diagram, add (57.5, 0) and (87.5, 0).

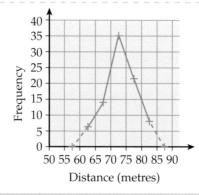

Distance (metres)

Exercise 8c

1 **a** Draw a histogram and frequency polygon for the data on
      the ages of members of a choral society.

Age, *x* years	$20 \leq x < 25$	$25 \leq x < 30$	$30 \leq x < 35$	$35 \leq x < 40$	$40 \leq x < 45$
Frequency	3	8	12	9	2

   b What is the modal class for ages in the choral society?

2 **a** Draw a frequency polygon for the data on the times taken
      to get to school by a group of students.

Time, *t* min	$0 \leq t < 10$	$10 \leq t < 20$	$20 \leq t < 30$	$30 \leq x < 40$	$40 \leq t < 50$
Frequency	27	18	6	4	1

   b Estimate the range of times students take to get to school.
   c What is the modal class for the time taken to get to school?

Problem solving

3 **a** Draw a frequency polygon for the data on the times taken for
      the 104 runners who completed the mens' 2008 USA Olympic
      marathon trial.

Time, *t* min	$120 \leq t < 130$	$130 \leq t < 140$	$140 \leq t < 150$	$150 \leq t < 160$	$160 \leq t < 170$
Frequency	1	38	49	15	1

   b What is the modal class for the time taken in the trial?
   c A number of runners did not complete the course.
      If they had, and their times were included in the graph,
      what differences would you expect to see?

4 In question **2**, there was only one student who took at least
   40 minutes to get to school.
   Looking at the shape of the distribution, do you think it is
   equally likely that the student took 40–45 minutes or
   45–50 minutes?
   Can you devise a better rule for estimating range for a set of
   data than using the maximum possible?

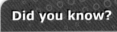

Did you know?

UK para-athlete David
Weir has won the
London Wheelchair
Marathon six times
and is the 2012
Olympic Champion.

8d Moving averages

Here is the index data for retail food sales for the period Spring 2006 to Autumn 2008.

p.106

2006				2007				2008		
Spr.	Sum.	Aut.	Win.	Spr.	Sum.	Aut.	Win.	Spr.	Sum.	Aut.
119.7	127.2	127.0	139.4	124.8	131.9	130.4	143.5	132.3	139.7	138.2

The short-term variation repeats in a **cycle** lasting four seasons or one year. To see the long-term trend it helps to smooth out the peaks and troughs during the cycle using a four-point **moving average**.

The first moving average is

$$\frac{119.7 + 127.2 + 127.0 + 139.4}{4} = \frac{513.3}{4} = 128.3$$ and it is plotted
in the centre of the values used in the calculation, halfway between Summer and Autumn 2006.

A line of best fit for the moving averages gives a **trend line**.

⦿ Using a trend line it is possible to predict future values of the moving average and individual data points.

Did you know?

The Retail Price Index (RPI) is a measure of the cost of a carefully selected set of items. It is compiled by the UK government each month to keep track of inflation.

Example

a Use the trend line for retail food sales to estimate the next moving average.
b Use your estimate to predict the index for Winter 2008.

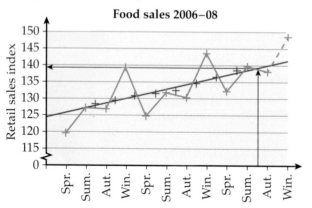

Food sales 2006–08

The further into the future you extrapolate (make predictions beyond the information given) the less reliable your predictions become.

a The next four-point average occurs halfway between Summer and Autumn 2008. Reading off from the trend line gives 139.
b Let x = the index in Winter 2008
$$\frac{(132.3 + 139.7 + 138.2 + x)}{4} = 139$$
$$410.2 + x = 4 \times 139$$
$$\Rightarrow x = 145.8$$

Exercise 8d

Problem solving

1   The table gives the numbers of orders of wedding cakes at Gina's
     bakery over a three-year period. The data is given quarterly.

Year	2006				2007				2008			
Quarter	Spr.	Sum.	Aut.	Win.	Spr.	Sum.	Aut.	Win.	Spr.	Sum.	Aut.	Win.
Number of cakes	15	30	22	10	19	35	24	11	20	37	27	14

a   Plot the data on a time-series graph.

b   Calculate four-point moving averages and
     plot them on the graph.

c   Draw a trend line for the moving averages.

d   Estimate the number of wedding cakes
     Gina's bakery will be asked for in Spring 2009.

The first moving average will be plotted on the horizontal axis halfway between Sum. and Aut. of 2006.

2   The table shows the number of units of electricity used by a
     household over a period of three years.

Year	2006				2007				2008			
Quarter	Spr.	Sum.	Aut.	Win.	Spr.	Sum.	Aut.	Win.	Spr.	Sum.	Aut.	Win.
Number of units	820	543	948	1185	835	527	931	1180	815	530	940	1162

a   Plot the data on a time-series graph.

b   Calculate four-point moving averages and plot them on the graph.

c   Draw a trend line for the moving averages.

d   Estimate the number of units that will be used in Spring 2009.

3   The table shows the number of workers in a factory who were not at
     work each day, because of illness or being on leave, over a three-week period.

Day	M	T	W	T	F	M	T	W	T	F	M	T	W	T	F
Number of absentees	12	5	4	3	11	10	4	5	5	12	11	4	2	6	10

a   Plot the data on a time-series graph.

b   Calculate five-point moving averages and plot them on the graph.

c   Is there a long term trend?

4   a   Why do you think the time series in questions **1** and **2** have the shapes that they do?

     b   In question **2**, if the data had been the cost of electricity in the same period, do you
          think the graph would have looked the same?

     c   In question **3**, how would you predict the number of workers not at work on the
          following Monday?

The **mean** is the total of the data values divided by the number of values.

With a **grouped frequency table**, which does not give the individual data values, you can only find an **estimated mean**.

▲ The mean house price in UK (2012) was £238 293.

Example

The table shows the lengths of shots of 85 golfers.

Distance, d	60 ≤ d < 65	65 ≤ d < 70	70 ≤ d < 75	75 ≤ d < 80	80 ≤ d < 85
Frequency	6	14	35	22	8

Calculate an estimate of the mean.

Multiply interval **mid-points** by frequencies.

$$6 \times 62.5 \qquad 14 \times 67.5 \qquad 35 \times 72.5 \qquad 22 \times 77.5 \qquad 8 \times 82.5$$
$$= 375 \qquad = 945 \qquad = 2537.5 \qquad = 1705 \qquad = 660$$

$$\frac{\text{estimate of total distance}}{\text{total shots}} = \frac{375 + 945 + 2537.5 + 1705 + 660}{6 + 14 + 35 + 22 + 8}$$

$$= \frac{6222.5}{85}$$

$$= 73.2058$$

$$= 73.2\,m \ (3\,sf)$$

Since you aren't given individual values assume the six shots in the first interval are all 62.5m – the interval's mid-point.

Calculating an estimated mean can be easier if you use an **assumed mean**. For the golf data assume a mean of 62.5 and subtract this from the mid-interval values. Then work out the mean of the new values and then add it back to the assumed mean.

Distance d	Frequency, f	Mid-interval, m	x = m − 62.5	f × x
60 ≤ d < 65	6	62.5	0	0
65 ≤ d < 70	14	67.5	5	70
70 ≤ d < 75	35	72.5	10	350
75 ≤ d < 80	22	77.5	15	330
80 ≤ d < 85	8	82.5	20	160
Total =	85		Total =	910

Estimated mean $= 62.5 + \dfrac{910}{85} = 62.5 + 10.7058$

$$= 73.2\,m \ (3\,sf)$$

Exercise 8e

1 The table shows the heights of a large sample of boys and girls.

Height, h cm	$130 \leq h < 140$	$140 \leq h < 150$	$150 \leq h < 160$	$160 \leq h < 170$	$170 \leq h < 180$
Number of boys	26	84	312	342	131
Number of girls	23	103	340	311	73

  a Without using an assumed mean, estimate both the mean heights of boys and of the girls in this sample.

  b What would be a sensible assumed mean to choose for this set of data?

2 The lengths of calls made by Naheed on her mobile phone are summarised in the table.

Time, t minutes	$0 \leq t < 1$	$1 \leq t < 3$	$3 \leq t < 5$	$5 \leq t < 10$	$10 \leq t < 20$
Number of calls	15	22	12	35	6

  a Estimate the mean length of Naheed's calls.

  b Would an assumed mean be helpful for this set of data? Give a reason for your answer.

3 The times taken for a group of Year 9 students to complete a set of three sudoku puzzles are recorded in the table.

Time, t minutes	$5 \leq t < 10$	$10 \leq t < 15$	$15 \leq t < 20$	$20 \leq t < 25$	$25 \leq t < 30$
Number of students	3	16	15	11	1

Using an assumed mean of 7.5 minutes, estimate the mean time taken to complete the puzzles.

4 The time taken for a group of students to complete a level in a computer game is shown.

Time, t minutes	$5 \leq t < 10$	$10 \leq t < 15$	$15 \leq t < 20$	$20 \leq t < 25$	$25 \leq t < 30$
Frequency	2	14	13	6	1

  a Find the modal class.

  b Find the class containing the median.

  c Find an estimate of the mean time.

Problem solving

5 In question **1**, if the heights had the last interval as $h \geq 170$ instead of $170 \leq h < 180$, how could you estimate the mean?

MyMaths.co.uk   Q 1201, 1254 **SEARCH**

A hypothesis that connects two variables often claims that the data will be **correlated**.

● Data shows a **linear correlation** if it tends to lie along a straight line in a scatter graph.

▶ A correlation is **positive/negative** if the **line of best fit** has a positive/negative gradient.

● The **strength** of a correlation measures how close the data is to the line of best fit.

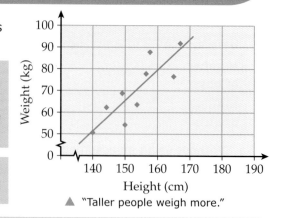

▲ "Taller people weigh more."

Example

The scatter graph shows the maximum distances at which a number of people could read a motorway sign and their ages.

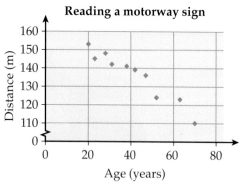

a Draw a **line of best fit** for this data.
b Use your line to **predict** the maximum distance at which someone aged 50 could read a motorway sign.

A line of best fit can be drawn by eye.

a

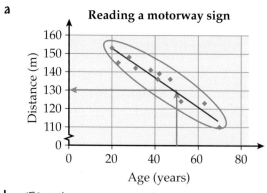

b 130 metres

It is only safe to make predictions for values within the range of data (interpolation).

Example

A new Head is considering altering the school's class sizes. He thinks that students in smaller classes obtain better exam results.

The scatter diagram shows the average GCSE score and the class size for last year's Year 11.

The Head says the current Year 10 should all be taught in classes of 30 to improve their chances at GCSE. Do you agree?

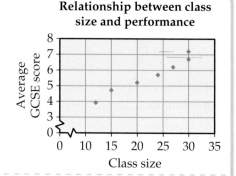

No, the direction of the cause and effect is the other way round.
The school has put the weaker academic pupils into smaller classes to offer them more support and putting them into a large class is likely to make their performance worse.

Exercise 8f

Problem solving

1 The scatter graph shows the age at marriage of a number of couples.
Use the line of best fit shown to estimate
a the age of the wife of a man aged 33
b the age of the husband of a woman aged 27.

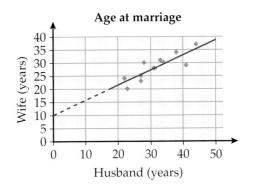

2 The scatter graph shows the prices of a number of cars of the same model and their ages.
a Explain why you should not draw a line of best fit.
b Could you predict what a 5-year-old car of this model would cost?

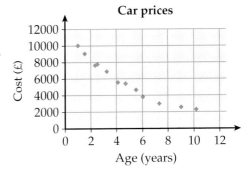

3 A geography student thinks that after a long eruption in a geyser there will be a longer wait for the next eruption to allow the pressure to build up. The graph shows the duration of an eruption, and the interval until the next eruption for 'Old Faithful'.
Describe the correlation.
Does this data support the student's claim?

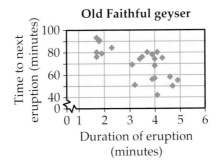

4 Lewis thinks that cars with smaller engines will have better fuel consumption. The graph shows the fuel consumption for a sample of new cars produced in May 2008.
a Does the data support Lewis' conjecture?
b Jensen says that heavier cars have bigger engines and it is actually the weight of the car which is important. That is, lighter cars use less fuel. Can you tell whether he is right from the graph?

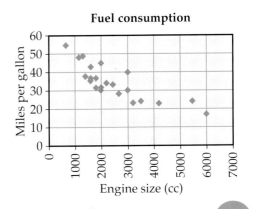

The range can be a poor measure of a data set's spread as it is affected by extreme values. A better measure of spread is based on **quartiles**.

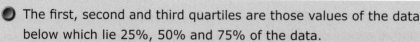

lowest value   median   highest value

lower quartile   upper quartile

- The first, second and third quartiles are those values of the data below which lie 25%, 50% and 75% of the data.
  ▶ The second quartile is also called the **median**.

- The **interquartile range (IQR)** is the difference between the third (**upper**) and first (**lower**) quartiles.
  ▶ The IQR measures the spread of the middle 50% of the data.

Quartiles can be read off a **cumulative frequency** diagram.

Example

The table shows normal pulse rates for some Year 9 students.

a Draw a cumulative frequency graph for this data.

b Estimate the
  i median
  ii upper and lower quartiles
  iii interquartile range.

Normal pulse rate, P	Frequency	Cumulative frequency	
$60 \le P < 70$	5	5	
$70 \le P < 80$	13	18	= 5 + 13
$80 \le P < 90$	30	48	= 18 + 30
$90 \le P < 100$	20	68	= 48 + 20
$100 \le P < 110$	12	80	= 68 + 12

a

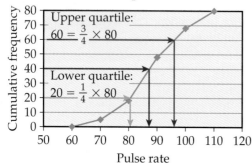

Upper quartile: $60 = \frac{3}{4} \times 80$

Lower quartile: $20 = \frac{1}{4} \times 80$

Add frequencies to find the cumulative frequency: 5 + 13 = 18, 18 + 30 = 48, and so on.

b i $\frac{1}{2} \times 80 = 40$, the 40th student has the median pulse rate.
  median   = 88   read off using the **blue** line.

ii upper quartile   = 96   read off using the **red** line.
  lower quartile   = 81   read off using the **green** line.

iii interquartile range = 96 – 81 = 15

The lower end of the first interval is 60, so plot (60, 0).

Exercise 8g

Problem solving

1 The table shows the numbers of people of different ages who were staying in a holiday resort during one particular week.

Age (x years)	0 ≤ x < 10	10 ≤ x < 20	20 ≤ x < 30	30 ≤ x < 40	40 ≤ x < 50
Number of people	15	8	35	12	7

Calculate the cumulative frequencies and use these to plot a cumulative frequency graph.

2 The cumulative frequency graph shows the number of characters in text messages sent by Alicia.

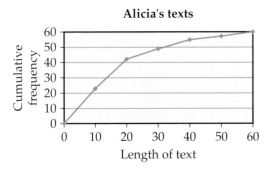

Cumulative frequency graphs are usually curves but can be drawn with line segments.

Estimate the median and the interquartile range of the text message lengths.

The intervals do not have to be the same size.

3 The table shows the amount of money a number of students raised for charity in a sponsored walk.

Sponsored (p £s)	0 ≤ p < 20	20 ≤ p < 30	30 ≤ p < 40	40 ≤ p < 50	50 ≤ p < 80
Number of people	12	18	13	10	7

a Draw a cumulative frequency graph.
b Estimate the median and the interquartile range of the amount of sponsorship raised.

4 The diagram shows some information about the weights of a group of boys and girls.
Orla says that the boys' graph is always to the right of the girls' graph, so all the girls are lighter than the boys. Is Orla right?

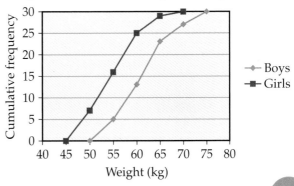

8h Interpreting data

When you interpret data, try to describe overall patterns.

Example

The table shows the percentages of boys and girls in England aged between 11 and 15 who said they had drunk alcohol in the week before the survey was taken.

	Boys				Girls			
	1994	1998	2002	2006	1994	1998	2002	2006
11	8	5	7	5	4	4	4	3
12	10	11	12	11	9	11	9	9
13	22	18	20	17	15	22	21	19
14	34	34	34	32	26	35	34	33
15	52	51	49	44	46	50	45	46

▲ Data from the *Health Information Service*.

a Draw two graphs to show this data for boys and girls.

b Is there any difference in the proportions of boys who drank alcohol for different ages?

c Is there any **trend** in the proportions of boys who drank alcohol over the period of these surveys?

d Compare the proportions of boys and girls who drank alcohol in these surveys.

A trend describes how the 'average' data value changes.

a

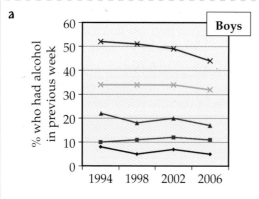

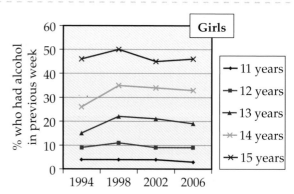

b The older the boys are, the higher the proportion who drank alcohol.

c There has been a decrease in the proportions at all ages except 12.

d As with boys, more older girls drink but there has not been the same general decrease in proportions over time.

By 2006 the proportion of 13-, 14- and 15-year-old girls drinking alcohol is higher than boys of the same age.

In 1994 there were more boys drinking than girls at all ages.

6

Exercise 8h

Problem solving

1 The table shows the percentages of adults in the UK (in 2007), who used the internet for each purpose.

	16–24	25–44	45–54	55–64	65+
Using email	84	87	86	86	80
Obtaining information from public authority websites	30	51	49	50	36
Internet banking	34	52	46	43	31
Playing/downloading games/images/films/music	58	40	25	14	-
Telephoning over the internet	15	13	10	-	-

▲ Data from the *Omnibus Survey, Office for National Statistics*.

a Draw a graph to show this data.

b Describe any differences between the different age groups.

2 The table shows the percentages of men and women of different age groups living alone in Great Britain in 1986/87 and in 2006.

	Men		Women	
	1986/87	**2006**	**1986/87**	**2006**
16–24	4.0	3.6	3.0	2.5
25–44	7.0	14.1	4.0	7.5
45–64	8.0	16.2	13.0	14.3
65–74	17.0	20.6	38.0	31.4
75 and over	24.0	32.1	61.0	61.3

▲ Data from the *General Household Survey (Longitudinal), Office for National Statistics (table 2.5 from Trends 38)*.

a Draw a graph to compare the proportions of men and women living alone in 2006.

b Describe any differences between the proportions of men and women living alone in 2006.

c Draw another graph to show the proportions in 1986/87 and comment on any changes there have been in the 20 years.

3 The graph show the proportions of 15-year-old boys and girls who had an alcoholic drink the previous week, for the surveys in 1994, 1998, 2002 and 2006.

Do you think this representation (with similar graphs for the other ages) makes it easier to make comparisons than the graphs used in the example?

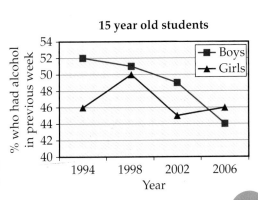

8i   Comparing distributions

● A **distribution** describes the way that data is spread out.
You can use measures of **average** and **spread** to compare distributions.

Things to compare

Average values	mode, median, mean
Spread	range, IQR
Shape	symmetric, skewed

When comparing **distributions** from graphs try to relate the comparisons to the context.

Example

The graph shows the time between eruptions for two geysers in Yellowstone National Park.
Give two similarities and two differences in the times between their eruptions.

Both distributions are have two distinct peaks and they have very similar spreads.
The range ≈ 1 hour.
The times between eruptions for Riverside are much greater than for Old Faithful by about 5 hours.

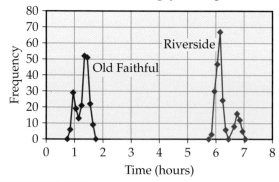

Times between geyser eruptions

Example

The graph shows the distances travelled by Arsenal and Aston Villa football clubs to Premier League away games during the 2008–09 season.
Compare the distances the teams have to travel for away games over the season.

Aston Villa have fewer matches that are only a short distance away than Arsenal, and they also have fewer matches that are a long distance away.
On average Arsenal have to travel further to their matches, with a median distance of 198 miles compared to 115 miles for Aston Villa and there is more variability in their journeys.

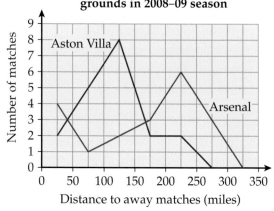

Distances from Arsenal and Aston Villa to other Premier League grounds in 2008–09 season

7

Exercise 8i

Problem solving

1 **a** What would you make of a newspaper headline which said this?

SUNDAY NEWS HOME

Mathematics is the easiest A-level!

b The two graphs show the number of candidates and the percentage of candidates getting each grade in the five most popular A level subjects. Give two features of the graphs which are consistent with the headline.

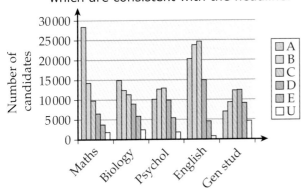

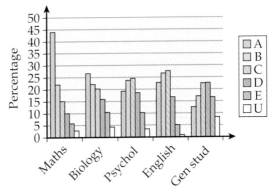

2 The graphs show a comparison between the weights of two samples of dunnocks (hedge sparrows) taken in January and in April. Compare the distributions of weights.

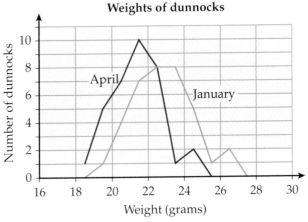

Weights of dunnocks

3 A transport company records the punctuality of its trains and buses over a week and the results of a random sample from each are summarised in the table (-5 means arrived 5 minutes early).

	$-10 \le t < 0$	$0 \le t < 10$	$10 \le t < 20$	$20 \le t < 30$	$30 \le t < 40$	$40 \le t < 50$
Number of buses	4	20	12	3	2	1
Number of trains	5	22	7	2	0	1

a On one graph, draw a frequency polygon for each type of transport.

b Compare the performances of the two types of transport.

4 The graph shows the fuel consumption and engine size of samples of new cars in 2002 and 2008.

Describe at least one similarity and at least one difference between the samples for 2002 and 2008.

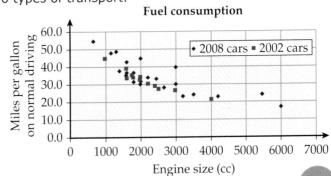

Fuel consumption

A **box plot**, also known as a box-and-whisker plot, is a diagram that summarises the location, spread and shape of a distribution.

Example

Construct a box plot to summarise this **cumulative frequency** data on the annual salaries of scientists.

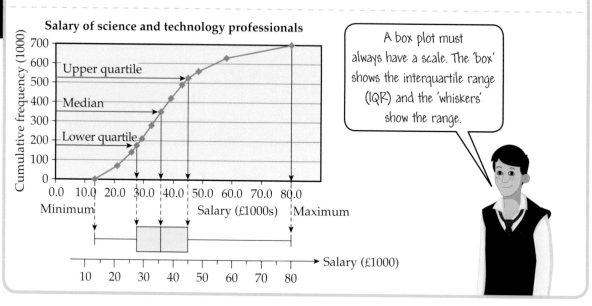

A box plot must always have a scale. The 'box' shows the interquartile range (IQR) and the 'whiskers' show the range.

Example

This box plot summarises the data on the salaries of graphic designers.

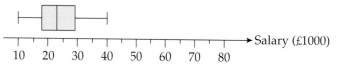

Comment on the difference in salaries between the graphic designers and scientists.

Always try to support your statements with statistics.

The typical salaries of designers are much less than those of scientists: median, £23k versus £36k.

The spread in salaries is also smaller for designers than scientists: IQR, 45 – 27 = £17k versus 28 – 18 = £10k.

For scientists, the long tail for high salaries indicates that some scientists earn significantly more than the average for their profession; this is not the case for designers.

Exercise 8j

1 For the height data summarised in the box plot what is the

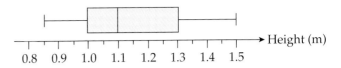

Height (m)

0.8 0.9 1.0 1.1 1.2 1.3 1.4 1.5

    a minimum   **b** median

    c range   **d** IQR.

2 Draw a box plot given the following information.

    a median = 25 cm   upper quartile = 31 cm   range = 27 cm

         minimum = 7 cm   IQR = 13 cm

    b median = 3.4   lower quartile = 2.8   range = 3.7

         maximum = 4.8   IQR = 1.1

Problem solving

3 Match the box plots to the graphs of their distributions.

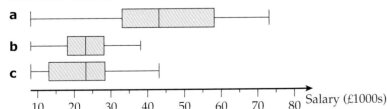

4 Compare the two pairs of distributions given their box plots.

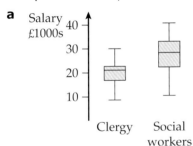

Did you know?

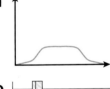

John Tukey developed the box plot in the 1970s. He is also credited with inventing the terms 'bit' and 'software'.

5 Given these box plots, sketch a bar chart for each possible data distribution.

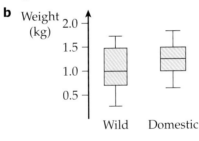

  10   20   30   40   50   60   70   80   Salary (£1000s)

Check out

You should now be able to ...

Test it ➡

Questions

✓	Draw a frequency polygon.	7	1
✓	Find trends using moving averages.	8	2
✓	Estimate the mean from a grouped frequency table.	7	3
✓	Interpret a scatter diagram.	7	4
✓	Draw and use a cumulative frequency graph.	8	5
✓	Compare distributions.	6	6
✓	Use box plots to make comparisons between data sets.	8	7

Language | Meaning | Example

Language	Meaning	Example
Hypothesis	A statistical statement to prove or disprove.	Small dogs live longer than large dogs.
Population	The whole group to be investigated.	All adults of voting age in UK
Sample	A small selection of the population.	A smaller sample chosen to predict the outcome of an election
Bias	Not fair.	A biased sample does not truly represent the whole population.
Frequency polygon	A graph showing the distribution of a grouped frequency table.	See page 136
Line of best fit	A straight line (or curve) drawn through a set of points to show a trend.	See page 142
Moving average	A way of removing season cycles to show the overall trend.	See page 138
Box plot	A diagram to show range, median and quartiles of a distribution.	See page 150
Quartiles	The value of the 75% (upper) and 25% (lower) item in an ordered data set.	See page 144
IQR	Interquartile range = upper quartile – lower quartile	See page 144

1. A group of students recorded how many pieces of homework they were set in a week.

Homework	0–4	5–9	10–14	15–19
Frequency	5	11	9	2

a. Estimate the range of the number of pieces of homework.

b. Find the modal class.

c. Draw a frequency polygon for the data.

2. The price of a share (pence) in a company over a 3-week period is shown in the table.

Week	1	2	3
Monday	25.9	26.0	34.8
Tuesday	25.7	26.1	35.7
Wednesday	25.9	26.3	38.0
Thursday	26.7	28.9	37.5
Friday	25.8	29.4	39.3

a. Plot the data on a time series graph.

b. Calculate the five-point moving averages and plot them on the graph.

c. Draw a trend line for the moving averages.

3. The number of words in text messages sent by Louise is summarised in the table.

Number of Words, w	Number of Messages
$0 < w \le 5$	20
$5 < w \le 10$	35
$10 < w \le 20$	15
$20 < w \le 40$	5

Estimate the mean number of words.

4. The table gives the percentage of the population of men and of who smoked cigarettes over several years.

Year	% Men	% Women
1974	52	41
1982	38	33
1990	31	29
1998	30	27
2006	26	23

Plot the percentages of men and women on a scatter graph and comment on any correlation.

5. For the data in question **3**.

a. Plot a cumulative frequency graph.

b. Use your graph to estimate the median.

c. Use your graph to estimate the IQR.

6. For the data in question **4**, draw a graph which shows the trend over time and compares men and women.

7. The table gives information about the age and gender of employees of a company.

Age	Male	Female
Median	34	29
Lowest	17	19
Highest	65	59
Lower Quartile	27	22
Upper Quartile	40	33

a. Draw a pair of box plots to show this information.

b. Use your box plots to compare the ages of male and female employees.

What next?

	Score		
Score	0 – 3		Your knowledge of this topic is still developing. To improve look at Formative test: 3C-8; MyMaths: 1194, 1195, 1201, 1204, 1213, 1248, 1249 and 1333
	4 – 6		You are gaining a secure knowledge of this topic. To improve look at InvisiPen: 413, 414, 427, 432, 433, 434 and 449
	7		You have mastered this topic. Well done, you are ready to progress!

MyMaths.co.uk

8a

1   For each of the following
    i   explain whether you should investigate it using an experiment or a survey
    ii describe how you would carry out the investigation
    iii estimate how long it would take you to collect the data.
    a   Do girls go to the cinema more often than boys?
    b   Does practice improve ability in being able to estimate a minute?

8b

2   A school wants to know how much exercise students take.
    The Deputy Head asks the head of PE to find out, so she surveys the first
    40 students who she sees playing games in the playground at lunchtime.
    a   Explain why this will give a biased sample and whether it is likely to over- or
            under-estimate the amount of exercise students take.
    b   Suggest a better way for the head of PE to take her sample.

8c

3   **a**   Draw a histogram and frequency polygon for the data on the weights
            of bags checked in on a flight.

Weight, x kg	$5 \leq x < 8$	$8 \leq x < 11$	$11 \leq x < 14$	$14 \leq x < 17$	$17 \leq x < 20$
Frequency	3	14	19	7	1

    b   What is the modal class for this data?
    c   Estimate the range of the weights of bags checked in.

8d

4   The table shows the number of days on which rain fell in a town
    over a period of three years.

Year	2006				2007				2008			
Quarter	Spr.	Sum.	Aut.	Win.	Spr.	Sum.	Aut.	Win.	Spr.	Sum.	Aut.	Win.
Number of days	33	33	29	32	36	34	28	33	37	35	30	33

    a   Plot the data on a time-series graph.
    b   Calculate four-point moving averages and plot them on the graph.
    c   Draw a trend line for the moving averages.
    d   Estimate the number of days rain fell in the town in Spring 2009.

8e

5   **a**   Estimate the mean weight of the bags checked in to the flight in question **3**.
    b   Would an assumed mean be helpful for this set of data?

6   The table shows the ages on their wedding day of the husband and wife for a sample of marriages at a church.

Husband	24	27	32	24	31	25	37	49	36	29
Wife	22	29	30	25	33	19	31	43	32	29

   a Draw a scatter diagram for this data and add a line of best fit.

   b The husband in another marriage at the church was 29.
Use your line of best fit to estimate the age of his wife.

7   The table shows the amount of money a number of events raised for a national charity in their centenary year.

Sponsorship(£1000s)	<10	$10 \leq x < 20$	$20 \leq x < 30$	$30 \leq x < 40$	$40 \leq x < 50$
Number of events	4	14	29	16	8

▲ Data from *Office for National Statistics* (*Social Trends* 38, table 6.9).

   a Draw a cumulative frequency graph.

   b Estimate the median and the interquartile range of the amount of money raised.

8   Describe the trend in rainfall from your answer to question **4**.

9   The chart shows a comparison of the weights of bags checked in on a long-haul and on a short-haul flight. Compare the distribution of weights of the bags on the two flights.

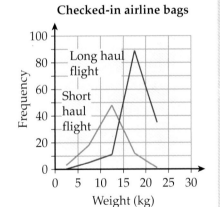

10 The boxplots are for the same data as in question **9**.

   a How much did the lightest bag on the long-haul flight weigh?

   b Amir says 'All the bags on the long-haul flight are about 4 kg more than the bags on the short-haul flight'. Explain what is wrong with Amir's claim.

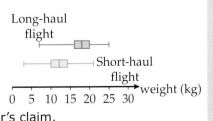

These questions will test you on your knowledge of the topics in chapters 5-8.
They give you practice in the types of questions that you may see in your GCSE exams.
There are 70 marks in total.

1   A rear car wiper blade has a length of 30 cm and covers an angle of 120°.
    Calculate in square metres the area of the rear window that is cleaned.   (2 marks)

2   O is the centre of a circle and the line AB
    is a tangent at P. The point Q is another
    point on the circle.

    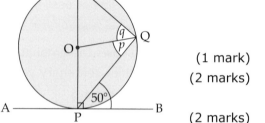

    a   What is the name of the line PQ?   (1 mark)
    b   Calculate the angle marked by the letter p.   (2 marks)
    c   Calculate the size of the angle marked by
        the letter q.   (2 marks)

3   Four regular pentagons are joined together
    as shown.

    a   Calculate angle x.   (2 marks)
    b   Calculate angle y.   (1 mark)

4   Are these statements true or false?
    a   The gradient of the line $2y = 4x - 1$ is 2.   (1 mark)
    b   The lines $y = 0$ and $x = 1$ are perpendicular.   (1 mark)
    c   The lines $y = 3x + 1$ and $2y = 6x + 1$ are parallel.   (1 mark)

5   a   Find the equation of the straight line in the form $y = mx + c$
        for a line that passes through the points (0, 3) and (-1, 5).   (3 marks)
    b   What is the equation of the line that is perpendicular to this
        line and passing through the point (0, 3)?   (3 marks)

6   a   Complete the table below for the cubic equation $y = x^3 + 2x - 1$.   (3 marks)

x	-3	-2	-1	0	1	2	3
y							

    b   Plot this graph on a suitable set of axes.   (3 marks)
    c   On the same graph plot the equation $y = 11x - 1$.   (2 marks)
    d   Give the coordinates of the points of intersection.   (2 marks)

7   a   Complete the table below for the equation $y = 3.7^{-x}$.   (2 marks)

x	-1	0	1	2	3	4
y						

    b   What type of function is $y = 3.7^{-x}$?   (1 mark)
    c   Plot the graph on a suitable set of axes.   (2 marks)
    d   What happens to the graph for larger and larger x values?   (1 mark)

8 The speed of light is 3.0×10^8 ms^{-1}.

  a The Sun is 148 800 000 km from the Earth. How long does it
   take light to reach us? (Give your answer in minutes.) (1 mark)

  b The nearest star Alpha Centauri is 4.367 light years away.
   How far away is this in kilometres? (2 marks)

9 Use your calculator to work out
 giving your answer to 2 dp.
 $$\frac{6.63 \times 10^{-34} \times 3.0 \times 10^8}{450 \times 10^{-9} \times 1.6 \times 10^{-19}}$$ (2 marks)

10 The table shows the number of monthly calls made by a mobile phone

108	82	52	60	75	102	95	53	67	82	140	155
J	F	M	A	M	J	J	A	S	O	N	D

  a Plot the data as a time-series graph. (4 marks)

  b Comment on the results obtained. (1 mark)

  c Calculate the five-point moving averages and plot these
   points on the same graph. (4 marks)

  d Comment on the trend. (1 mark)

11 The table shows how the fuel consumption (in litres per 100 km)
 changes as the speed of the car increases.

Speed (kph)	20	30	40	50	60	70	80
Fuel consumption (litres/100 km)	9.8	8.8	8.0	6.9	6.1	5.6	4.5

  a Plot a scatter graph for this data. (3 marks)

  b Draw the line of best fit. (1 mark)

  c Comment on the type of correlation obtained. (2 marks)

12 The frequency table gives the heights of 50 boys and 50 girls

Height h (cm)	Frequency (boys)	Frequency (girls)	Cumulative frequency (boys)	Cumulative frequency (girls)
$155 \leq h < 160$	2	4		
$160 \leq h < 165$	5	12		
$165 \leq h < 170$	15	22		
$170 \leq h < 175$	20	8		
$175 \leq h < 180$	7	4		
$180 \leq h < 185$	1	0		

  a Complete the cumulative frequency table. (3 marks)

  b For each distribution, construct a
   i cumulative frequency diagram (4 marks)
   ii box plot. (4 marks)

  c Estimate the median height for boys and girls and comment on the results. (3 marks)

9 Transformations and scale

Introduction

Since the 15th century artists have used the ideas of perspective to draw pictures of three dimensional objects and images. The use of constructional lines which meet at a 'vanishing point' allows the artist to accurately draw and position smaller objects so that they appear to be in the distance and therefore give the illusion of depth to the picture.

What's the point?

The construction lines and vanishing points used by artists are based upon the mathematics used to construct enlargements.

Objectives

By the end of this chapter, you will have learned how to ...

- ◖ Reflect, rotate and translate 2D shapes.
- ◖ Enlarge 2D shapes using positive and negative scale factors.
- ◖ Use and interpret maps and scale drawings.
- ◖ Calculate unknown lengths in similar shapes.

Check in

1 **a** Copy the diagram on square grid paper and draw the
       flag after a reflection in the mirror line.
   b Translate this image 3 units to the right and 1 unit up.
       Draw this new image on your diagram.

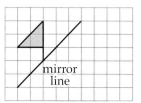

2 A scale drawing of a door has a scale of 1 cm to
   represent 50 cm. The door measures 3.9 cm by
   1.6 cm on the scale drawing.
   Calculate the actual height and width of the door.

Scale: 1 cm represents 50 cm

Starter problem

Look at this photograph. You can see some lines
which are parallel and some which seem to
converge at a point called the vanishing point.

Investigate the heights of the 'objects in the
photograph' and their distances from the
vanishing point.

9a Transformations

You can transform 2D shapes using a

reflection (flip)   **rotation** (turn)   **translation** (slide).

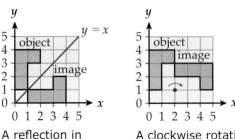

A reflection in the line $y \cdot x$

A clockwise rotation of 90° about (2, 1)

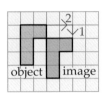

A translation of $\begin{pmatrix} 2 \\ -1 \end{pmatrix}$

$\begin{pmatrix} 2 \\ -1 \end{pmatrix}$ means 2 to the right and 1 down.

$\begin{pmatrix} -1 \\ 3 \end{pmatrix}$ means 1 to the left and 3 up.

The **transformation maps** the **object** to the **image**.

The object and the image are **congruent** in reflections, rotations and translations.

Congruent figures are the same shape and the same size.

You can transform 2D shapes using a sequence of transformations.

Example

A and B are two transformations.
▶ A is a reflection in the line $y = x$
▶ B is a rotation of 180° about (0, 0)
One of the transformations is applied to the pink triangle and then the other transformation is applied to its image.
Does the order matter in which these transformations are applied?

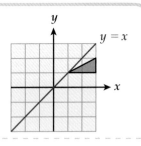

A then B

Pink $\xrightarrow{\text{Reflection}}$ Blue $\xrightarrow{\text{Rotation}}$ Green

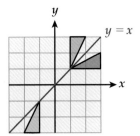

B then A

Pink $\xrightarrow{\text{Rotation}}$ Blue $\xrightarrow{\text{Reflection}}$ Green

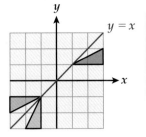

If the order of the transformations does not matter, then they are **commutative**. This is not always the case!

In this example, the order of the transformations does not matter.

Exercise 9a

Problem solving

1 a Copy the diagram on square grid paper.
Reflect the green triangle in the line $x = 2$ to give the image I_1.

b State the coordinates of the vertices of triangle I_1.

c Reflect triangle I_1 in the line $y = 2$ to give the image I_2.

d State the coordinates of the vertices of triangle I_2.

e Describe fully the single transformation that maps the green triangle to triangle I_2.

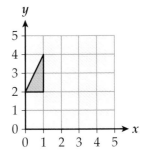

2 A and B are two transformations.

A is a translation of $\begin{pmatrix} -1 \\ 3 \end{pmatrix}$. 　　B is a translation of $\begin{pmatrix} 3 \\ -2 \end{pmatrix}$.

Find the single transformation that is equivalent to translation A followed by translation B.

3 A and B are two transformations.

▶ A is a reflection in the x-axis.

▶ B is an anticlockwise rotation of $90°$ about $(0, 0)$.

The pink triangle has coordinates $(1, 1)$, $(3, 1)$ and $(3, 2)$.

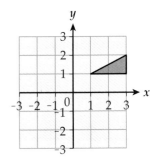

a On two separate diagrams draw the following two sequences of transformations, labelling the images I_1, I_2 and I_3, I_4.

Pink $\xrightarrow{\text{A}} I_1 \xrightarrow{\text{B}} I_2$ 　　Pink $\xrightarrow{\text{B}} I_3 \xrightarrow{\text{A}} I_4$

b Give the coordinates of the images I_1, I_2, I_3 and I_4.

c Does the order in which these transformations are applied matter?

4 What single transformation is equivalent to the following pairs of transformations?

Look at the effect of the transformations on a 'flag' shape.

a Reflection in $x = 2$ followed by reflection in $x = 4$.

b Reflection in $y = 1$ followed by reflection in $y = -1$.

c Reflection in $x = 1$ followed by reflection in $y = x$.

d Reflection in $y = 2$ followed by rotation through $180°$ about $(0, 2)$.

⬤ An **enlargement** is a transformation that alters the size of a shape. The angles of the shape do not change.
　▶ Under an enlargement the object and image are **similar** shapes.

p.268 ▶ You enlarge a shape by multiplying the lengths of the shape by the **scale factor**. The position of the image is fixed if you use a **centre of enlargement**.

Draw lines from the centre of enlargement through the vertices of the object.

Always measure from the centre of enlargement.

Distance to image vertex = scale factor × distance to object vertex.

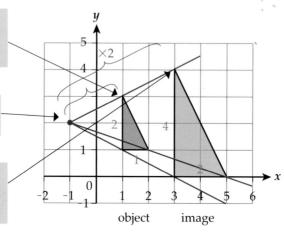

object　　image

The enlargement is scale factor 2, using (-1, 2) as the centre of enlargement.

Example

a Using the given centre, enlarge the pink rectangle by scale factor $\frac{1}{3}$.

b Calculate the perimeter and area of the pink rectangle and the enlargement.

Centre of enlargement

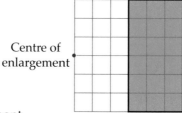

If the scale factor is positive and <1, the image is smaller than the object.

a

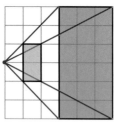

All the lengths have been multiplied by $\frac{1}{3}$.

b Perimeter of the pink rectangle = 18 units
　Perimeter of the enlargement = 6 units
　　Ratio of the perimeters = 3 : 1
　　　　　　　　　　　　 = 1 : SF
　Area of the pink rectangle = 18 square units
　Area of the enlargement = 2 square units
　　Ratio of the areas = 9 : 1
　　　　　　　　　 = 1 : (SF)²

Measure the ratio of image and object sides to check the scale factor.
$1 \div 3 = \frac{1}{3}, 2 \div 6 = \frac{1}{3}.$

Exercise 9b

1 a Calculate the perimeters and areas of the twelve pentominoes.

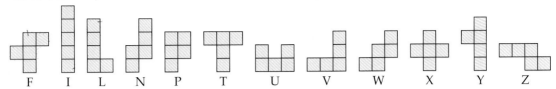

F   I   L   N   P   T   U   V   W   X   Y   Z

Each pentomino is enlarged by scale factor 2.

b Calculate the perimeter and area of each enlarged shape.

2 Draw the enlargement of each shape using the dot as the centre of enlargement and the given scale factor.

a
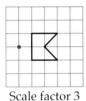

Scale factor 3

b
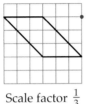

Scale factor $\frac{1}{3}$

c

Scale factor $\frac{1}{2}$

Problem solving

3 In each diagram, the pink shape is an enlargement of the green shape. Copy each of these shapes onto coordinate axes to calculate the scale factor and to find the coordinates of the centre of enlargement.

a
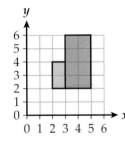

b
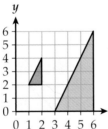

c
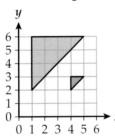

4 This triangle is enlarged by scale factor p.
The image is then enlarged by scale factor q.
What is the scale factor of the single enlargement that is equivalent to these successive transformations?

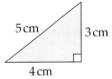

5 cm   3 cm   4 cm

5 Fold a sheet of A4 paper in half. The size of the paper is now A5.
Folding a sheet of A5 paper gives size A6 and so on.
A sheet of A4 paper measures 297 mm by 210 mm.
The different sizes of paper can be arranged with opposite corners along a diagonal, so that A4 is an enlargement of A5 *etc.*
Find the scale factor of the enlargement.

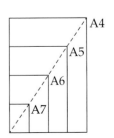

A4
A5
A6
A7

You can enlarge a shape with a **negative scale factor**.

● If the scale factor is negative, the image and the object are on opposite sides of the centre of enlargement.

The object and the image are similar
- the lengths increase in proportion
- the angles stay the same.

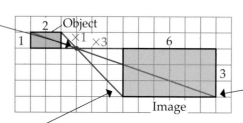

Centre of enlargement

Image

Object

The scale factor is -2

Always measure from the centre of enlargement.

Draw lines from the centre of enlargement to the vertices of the object.

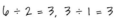

Distance to image vertex = scale factor × distance to object vertex.

The enlargement is scale factor -3
The lengths have been multiplied by 3

Measure the ratio of image and object sides to check the scale factor.
$6 \div 2 = 3$, $3 \div 1 = 3$

Example

The green triangle is an enlargement of the orange triangle.

Find

a   the centre of enlargement

b   the scale factor.

a   Join the corresponding vertices to find the centre of enlargement.

b   Compare corresponding lengths of the triangles to find the scale factor and insert the negative sign.

Scale factor = $\frac{-6}{3}$ = -2

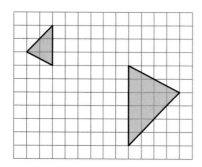

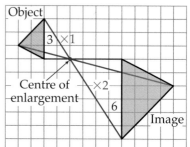

Object

Centre of enlargement

Image

Exercise 9c

1   Draw the enlargement of each shape using the dot as
the centre of enlargement and the given scale factor.

a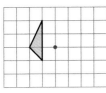

Scale factor -2

b

Scale factor -1

c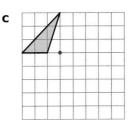

Scale factor -3

2   Each pink shape is an enlargement of the blue shape.
Copy these shapes onto coordinate axes to calculate the scale
factors and find the coordinates of the centres of enlargement.

a 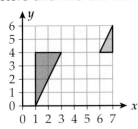   **b** 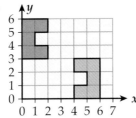   **c**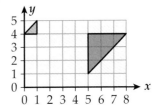

Problem solving

3   Draw triangle ABC on a coordinate grid.
  a Write down the coordinates of the points A, B and C.
  b Enlarge the triangle by scale factor -2 using (0, 0) as the
centre of enlargement.
Label the image A′, B′ and C′.
  c Write down the coordinates of the points A′, B′ and C′.
  d What do you notice about the coordinates of the
vertices of the object and the image?

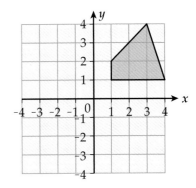

4 **a** Enlarge the green quadrilateral by scale factor -1
using (0, 0) as the centre of enlargement.
  b Describe a single different transformation that is
equivalent to this enlargement.
  c Is this enlargement still equivalent to your
transformation if the object is in a different quadrant
of the coordinate grid?

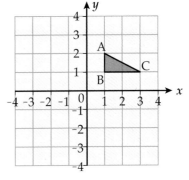

You use **maps** and **scale drawings** to represent real-life distances.

- Real-life distances are reduced in proportion using a **scale**.
- All the angles remain the same.
- The scale enables you to interpret the map or scale drawing.

⬤ A scale can be written as a **ratio**.

In a map or scale drawing, corresponding lengths
are in the same ratio.

◀—————30 metres—————▶

▲ The length of the blue
whale is 30 metres.

◀—————6 cm—————▶

▲ Scale 1 : 500

The real-life lengths are an enlargement, scale factor 500, of the
scale drawing.

The lengths in the scale drawing are 500 times smaller than in real life.

30 m ÷ 500 = 3000 cm ÷ 500 = 6 cm

Example

The distance from Leeds to York is 40 km.

The distance from Leeds and York on
the map is 2.5 cm.

Express the scale of the map as

a 1 cm represents _____

b as a ratio in the form 1 : n.

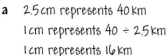

York

2.5 cm

Leeds

a 2.5 cm represents 40 km

   1 cm represents 40 ÷ 2.5 km

   1 cm represents 16 km

b Scale is 2.5 cm : 40 km

   2.5 cm : 40 000 m

   2.5 cm : 4 000 000 cm

   1 : 1 600 000

   The real-life distances are
   1 600 000 times longer.

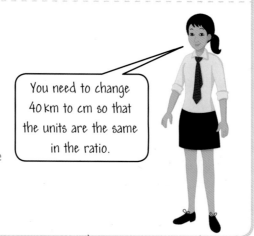

You need to change
40 km to cm so that
the units are the same
in the ratio.

Did you know?

The Gough map is
one of the earliest
maps of Britain. It
was drawn around
1370 and is kept in
the Bodleian library
in Oxford.

6

Exercise 9d

1 A map has a scale of 1 : 25 000.
   Calculate the real-life distance in kilometres that these lengths represent.
   a 4 cm   **b** 0.5 cm
   c 8.5 cm   **d** 25 cm
   e 35 cm   **f** 2 mm

2 Calculate the distance in centimetres on the map for these real-life distances.
   a 5 km   **b** 7.5 km
   c 4.4 km   **d** 10.5 km
   e 0.5 km   **f** 200 m

3 Write each of these scales in the form 1 : *n*.
   a 1 cm represents 1 m
   b 1 cm represents 5 m
   c 1 cm represents 0.25 m
   d 4 cm represents 1 km
   e 10 cm represents 5 km
   f 1 cm represents 0.25 km
   g 5 mm represents 10 m
   h 10 cm represents 1 mm

Problem solving

4 A London Routemaster bus has these dimensions
       length 9.14 m   height 4.37 m   width 2.44 m.
   A scale model is made using a scale of 1 : 50.
   Calculate the length, height and width of the model in centimetres, giving each answer to the nearest millimetre.

5 The centre of a dartboard must be 1.73 m from the floor.
   A darts player must stand behind the 'oche', which is 2.37 m from the wall.
   Use this information to
   a draw a scale drawing using a scale of 1 : 25
   b measure and calculate the distance from the oche to the centre of the dartboard.

Dartboard

1.73 m

Oche ←——— 2.37 m ———→

6 The Highway Code gives the distances in metres a driver needs to stop a vehicle at various speeds.
   Each stopping distance is calculated by adding the thinking distance and the braking distance.
   a Draw a poster to illustrate the thinking distance, the braking distance and the stopping distance for each speed.
      Use a scale of 1 : 500 for the distances.
   A typical car can be 5 metres long.
   b Superimpose cars on the distances on your scale drawing.

Speed (mph)	Thinking distance (metres)	Braking distance (metres)
20	6	6
30	9	14
40	12	24
50	15	38
60	18	55
70	21	75

9e Similar shapes

In an enlargement, the object and the image are **similar**.

– The lengths increase in proportion.
– The angles stay the same.

You can use corresponding lengths to find the scale factor of an enlargement.

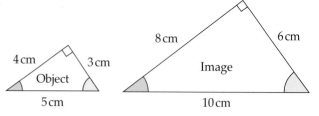

> ● Scale factor = $\dfrac{\text{length on the image}}{\text{corresponding length on the object}}$

scale factor = $\dfrac{10}{5}$ = 2

scale factor = $\dfrac{6}{3}$ = 2

scale factor = $\dfrac{8}{4}$ = 2

Example

Calculate the length a in these similar triangles.

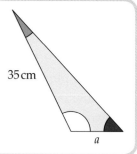

Scale factor = $\dfrac{35}{10}$ = 3.5   Compare corresponding sides.

a = 7 × 3.5 = 24.5 cm

You can use similar triangles to divide a line in a given **ratio**.

Example

A and B have coordinates (5, 10) and (25, 20) respectively.
Find the coordinates of the point P that divides the line
AB in the ratio 2 : 3.

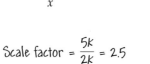

Extract two similar triangles from the grid.

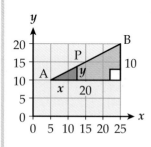

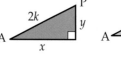

AP = $\dfrac{2}{5}$ × AB

Scale factor = $\dfrac{5k}{2k}$ = 2.5

x = 20 ÷ 2.5 = 8

y = 10 ÷ 2.5 = 4

The coordinates of P are (5 + 8, 10 + 4) or (13, 14).

Exercise 9e

1 Each pink rectangle is an enlargement of the blue rectangle.
Calculate the scale factor of each enlargement.

a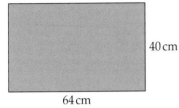
24 cm
15 cm

b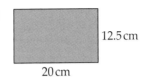
4 cm
2.5 cm

8 cm
5 cm

c
64 cm
40 cm

d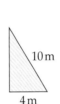
20 cm
12.5 cm

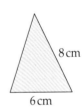
2 Calculate the scale factor and the missing length in each pair of similar triangles.

a
2 cm
1 cm
a
3 cm

b
10 m
4 m
b
6 m

c
8 cm
6 cm
c
1.5 cm

Problem solving

3 A and B have coordinates (2, 3) and (18, 15) respectively.
   a Find the coordinates of the point P that divides the line AB in the ratio 1 : 3.
   b Find the coordinates of the midpoint of AB.

4 A pink cuboid has dimensions 3 cm by 4 cm by 5 cm.
   The lengths of the pink cuboid are enlarged by scale factor 3 to give the blue cuboid, scale factor 4 to give the green cuboid and scale factor k to give an orange cuboid.
   Calculate the ratio of
      i lengths   ii surface areas   iii volumes
   a for the pink and blue cuboids
   b for the pink and green cuboids
   c for the pink and orange cuboids.

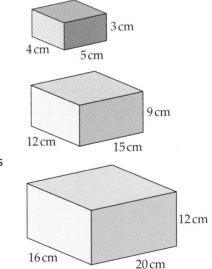
3 cm
4 cm
5 cm
9 cm
12 cm
15 cm
12 cm
16 cm
20 cm

Check out

You should now be able to ...

Test it ➡

Questions

✓ Reflect, rotate and translate 2D shapes.	6	1, 2
✓ Enlarge 2D shapes using positive and negative scale factors.	7	3 – 5
✓ Use and interpret maps and scale drawings.	6	6
✓ Calculate unknown lengths in similar shapes.	7	7

Language Meaning Example

Language	Meaning	Example
Congruent	Shapes that are the same shape and size.	Two squares with sides 3 cm are congruent
Similar	Shapes that are the same shape but not the same size.	All equilateral triangles are similar.
Centre of enlargement	A point used to set the position of the image after an enlargement.	
Scale factor	The number used to calculate the position and lengths of the image in an enlargement.	An enlargement scale factor two gives an image with sides 2 × those of the object.
Scale drawing	An accurate drawing of an object using a ratio.	A map uses 4 cm to represent 1 km, or 1 : 25 000

1 On a set of axes, both from -8 to 8, draw a triangle with vertices (-8, -7), (-5, -1) and (-2, -5).
Label the triangle A.

 a Rotate shape A 90° anticlockwise about (0, 0) and label the image B.

 b Reflect B in the line $y = 0$ and label the image C.

 c Rotate C 90° anticlockwise and label the image D.

 d Describe fully the single transformation that moves A to D.

2 A is a translation of $\begin{pmatrix} 2 \\ -5 \end{pmatrix}$.

 B is a translation of $\begin{pmatrix} -1 \\ -2 \end{pmatrix}$.

 a Find the single transformation that is equivalent to translation A followed by translation B.

 b Find the image of the point (-9, 3) under this single transformation.

3 Enlarge the shape by scale factor $\frac{1}{2}$ using the dot as the centre of enlargement.

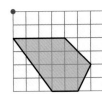

4 B is enlarged to give B'. Calculate the scale factor and find the coordinates of the centre of enlargement.

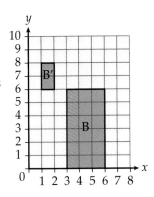

5 Enlarge this shape by scale factor -2 using the dot as the centre of enlargement.

6 John starts at A, walks due north for 3.36 km then east for 4.4 km and arrives at B.

 a Draw a scale drawing of his journey using a scale of 1 : 80 000

 b How far his John from his starting point?

7 Calculate the missing length in this pair of similar triangles.

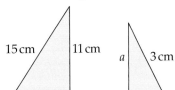

What next?

Score			
	0 – 3		Your knowledge of this topic is still developing. To improve look at Formative test: 3C-9; MyMaths: 1125, 1099, 1117 and 1119
	4 – 6		You are gaining a secure knowledge of this topic. To improve look at InvisiPen: 317, 366, 368 and 372
	7		You have mastered this topic. Well done, you are ready to progress!

1 A and B are two transformations.

▶ A is a reflection in the line $y = x$
▶ B is a reflection in the line $y = -x$

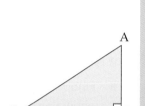

The green triangle has vertices (1, 1), (3, 1) and (3, 2).

a On two diagrams draw the following two sequences of transformations, labelling the images I_1, I_2 and I_3, I_4.

$$\text{Green} \xrightarrow{\text{A}} I_1 \xrightarrow{\text{B}} I_2 \qquad \text{Green} \xrightarrow{\text{B}} I_3 \xrightarrow{\text{A}} I_4$$

b Give the coordinates of the images I_1, I_2, I_3 and I_4.

c Does the order matter in which these transformations are applied?

2 This right-angled triangle is reflected in the line AB.

The triangle ABC and its image are now reflected in the line CB extended.
State, with reasons, the mathematical name of the large shape formed by the object and the images.

3 Calculate the dimensions of the photograph after these enlargements.

a 20" by 30" enlarged by scale factor $\frac{1}{5}$

b 16" by 24" enlarged by scale factor $\frac{1}{2}$

c 16" by 24" enlarged by scale factor $\frac{1}{4}$

> **Hint:** " means inches.

4 Draw the enlargement of each shape using the dot as the centre of enlargement and the given scale factor.

a

Scale factor 3

b

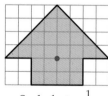

Scale factor $\frac{1}{2}$

c

Scale factor $\frac{1}{3}$

5 Draw the enlargement of each shape using the dot as the centre of enlargement and the given scale factor.

a

Scale factor -2

b

Scale factor -1

c

Scale factor -2

d

Scale factor -1

6 Write each of these scales in the form 1 : *n*

  a 1 cm represents 0.2 m   **b** 1 cm represents 25 m
  c 1 cm represents 2.5 km   **d** 1 cm represents 50 m
  e 5 cm represents 10 km   **f** 2 cm represents 250 m

7 A velociraptor was a dinosaur measuring about 2 metres long and about 0.5 metres high. Calculate the length and height of this dinosaur in a scale drawing, if the scale is

  a 1 : 20   **b** 1 : 5   **c** 1 : 50   **d** 1 : 25   **e** 1 : 4.

8 Calculate the scale factor and the missing length in each pair of similar triangles.

a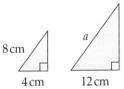

8 cm   *a*
4 cm   12 cm

b

20 cm   *b*
15 cm   3 cm

c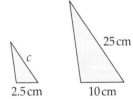

25 cm   *c*
2.5 cm   10 cm

9 A and B have coordinates (1, 10) and (10, 4) respectively.
Find the coordinates of the point P that divides the line AB in the ratio

  a 1 : 2
  b 2 : 1

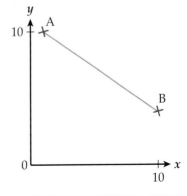

The Earth's climate has always changed due to natural causes such as change of orbit, volcanic eruptions and changes in the sun's energy, but now there is real concern that human activity is upsetting the balance by adding to 'greenhouse gases'.

Greenhouse gases

This diagram shows some of the main factors behind global warming.

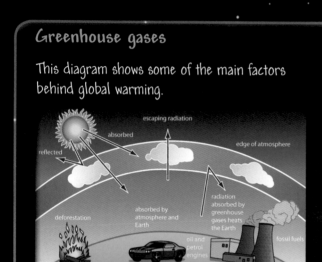

[Task 1]

The pie chart shows the contribution made to the warming effect by the main greenhouse gases.

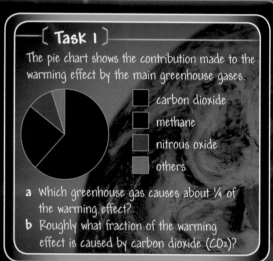

☐ carbon dioxide

☐ methane

☐ nitrous oxide

☐ others

a Which greenhouse gas causes about ¼ of the warming effect?

b Roughly what fraction of the warming effect is caused by carbon dioxide (CO_2)?

[Task 2]

THE DAILY NEWS

What's News ▷

GREENHOUSE GASES UP BY 25%

A recent report says that in the last 60 years, CO_2 concentrations have increased by around 25%.

◀ ◯ ▶

a Does the graph show the same increase in CO_2 levels as the report? Show your workings out.

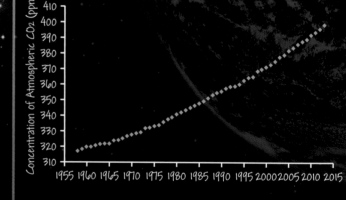

b In the year 1000 AD, CO_2 concentration was around 280 ppm (parts per million). By what percentage has it increased?

c (Harder) If CO_2 concentrations continue at their current rate, in what year would you predict the CO_2 concentration to reach 500 ppm?

Global temperature change

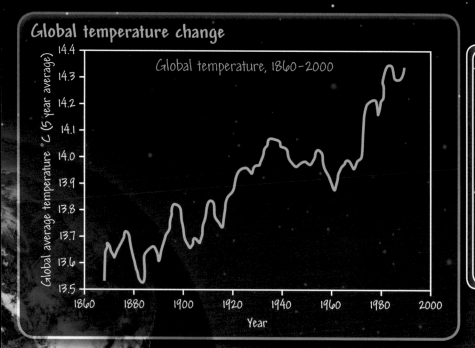

Global temperature, 1860–2000

Global average temperature °C (5 year average) vs *Year*

[Task 3]

a Describe in words what the graph shows.

b Estimate the global temperature in
  i 1880
  ii 1980

c Calculate the percentage change in temperature, giving the answer on your calculator display to 1 decimal place.

[Task 4]

Here are the monthly temperatures for Oxford in 1908 and 2008.

1908	Jan	Feb	Mar	Apr	May	June	July	Aug	Sep	Oct	Nov	Dec
max °C	5.5	8.6	8.3	10.4	17.6	20.5	21.5	20	17	16.1	11.1	6.5
min °C	-0.4	2.7	1.5	2.4	9	9.8	12.2	10.7	9.4	8.1	5	1.6

2008	Jan	Feb	Mar	Apr	May	June	July	Aug	Sep	Oct	Nov	Dec
max °C	10.3	10.5	10.6	13.1	18.7	20.1	21.7	20.8	17.6	14.3	9.9	6.5
min °C	4.7	1.7	3.5	4.7	9.5	11.1	12.8	13.9	10	6.3	4.6	1.3

a Write down the range in
  i maximum temperatures,
  ii minimum temperatures, for each year.

b Calculate the mean of
  i maximum temperatures,
  ii minimum temperatures, for each year.

c Use your results from a and b to compare the temperatures in Oxford in 1908 and 2008. Are your findings in agreement with the graph in Task 3?

d Why might using data for just two years in a single city not be adequate to make any long-term conclusions about global warming?

10 Equations

Introduction

You might think that equations exist only in maths textbooks, but …

… Astronomers use equations to describe the movements of planets and the paths of asteroids. Nuclear physicists use equations to calculate the half-lives of dangerous radioactive isotopes. Biologists use equations to predict the likely growth or decline of populations of endangered animals. The police force use equations to determine the cause of accidents by calculating the speeds and braking distances of the vehicles involved.

What's the point?

Equations are used in all walks of life to model complex real-life situations.

Objectives

By the end of this chapter, you will have learned how to …

- Solve linear equations with brackets and algebraic fractions.
- Solve simultaneous equations by elimination.
- Solve simultaneous equations by drawing graphs.
- Solve linear inequalities with one variable.
- Find approximate solutions to equations using trial-and-improvement.

Check in

1   Expand and simplify these expressions.

   a $12(2x + 10)$ **b** $3(4a + 1) + 8(a + 3)$

   c $5x(x - 4)$ **d** $3(2y + 5) - 4(5 - y)$

2   Solve these equations.

   a $5x + 15 = 30$ **b** $3(2m - 4) = 30$

3   All of these expressions except one have a value of 13 when $x = 2$.
    Which is the odd one out?

   $3x^2 + 1$ $x^3 + 5$ $x^2 + x + 5$ $20 - 2x^2 + 1$

Starter problem

Jake draws a 4×4 square grid. He says that there are
16 squares in the grid. He is wrong. How many squares
are there on a 4×4 grid?

How many squares are there on an 8×8 chessboard?

How many squares are there on an $n \times n$ chessboard?

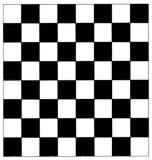

10a Consolidating linear equations

> Linear equations with unknowns on both sides can be solved
> by subtracting the smallest algebraic **term** from each side.

Example

The areas of carpet
needed to cover
two bedrooms in a
house are equal.
Find the dimensions of each room.

8

4

$8 - x$

$x + 1$

You may have to
construct an equation
from given information.

Area of rectangle = length × width

$8(8 - x) = 4(x + 1)$

$64 - 8x = 4x + 4$

$64 = 12x + 4$

$60 = 12x$

$x = 5$

-8x is smaller than 4x so subtract
-8x from both sides. This is the
same as adding 8x to both sides.

The first room measures 3 m by 8 m and the second room 4 m by 6 m.

> If an equation has just one negative algebraic term you can add
> the inverse of the negative term to both sides.

Example

Solve $10 - 5x = 12$

$10 - 5x = 12$

$10 = 12 + 5x$ Add 5x to both sides.

$-2 = 5x$

$x = -\dfrac{2}{5}$ Divide both sides by 5.

Solutions can be fractions,
negative or both. Fractions
are exact and you don't
need to convert to decimals.

> Equations with fractions can often be solved using **cross-multiplication**.

Example

Solve

$\dfrac{4}{2 - 3x} = \dfrac{5}{6 - 2x}$

$\dfrac{4}{2 - 3x} = \dfrac{5}{6 - 2x}$

$4(6 - 2x) = 5(2 - 3x)$

$24 - 8x = 10 - 15x$

$24 + 7x = 10$

$7x = -14$

$x = -2$

Just as $\left[\dfrac{10}{2} \times \dfrac{25}{5}\right]$ leads to

$10 \times 5 = 25 \times 2$

then $\left[\dfrac{4}{2 - 3x} \times \dfrac{5}{6 - 2x}\right]$ leads to

$4(6 - 2x) = 5(2 - 3x)$.

Exercise 10a

1 Solve these equations.

a $4x + 7 = 19$ b $7 - 2x = -3$ c $\dfrac{x + 5}{3} = 5$

d $4(x + 1) = 4$ e $3x + 6 = 2x + 9$ f $4x - 5 = 10 - x$

g $2(7x + 1) = 3(x + 8)$ h $12 - x = 4(x + 13)$ i $\dfrac{2x + 7}{3} = x + 1$

2 Solve these equations.

You should find your solution in the box.

a $5x - 12 = -2$ b $12 - 3y = 2$ c $2(4x + 1) = 5(3x - 2)$

d $6 - 4y = 10 - 8y$ e $100 - a = 56$ f $\dfrac{x + 12}{5} + 3 = 5$

g $\dfrac{y + 1}{6} = \dfrac{3y}{5}$ h $\dfrac{3}{x + 5} = \dfrac{9}{2x + 1}$ i $\dfrac{2(3x - 1)}{4} = \dfrac{x + 9}{3}$

| A 1 | B -2 | C $\dfrac{5}{13}$ | D 2 | E 44 | F $3\frac{1}{3}$ | G -14 | H $1\frac{5}{7}$ | I 3 |

Problem solving

3 Amie solved three equations but got them all wrong.

Can you spot and correct her error in each case?

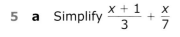

a $8y - 1 = 4y + 1$	b $18 - 3a = 6 - 9a$	c $\dfrac{x + 6}{2} = \dfrac{x + 14}{3}$
$4y - 1 = 1$	$18 = 6 - 12a$	$3x + 6 = 2x + 14$
$4y = -2$	$12 = 12a$	$x + 6 = 14$
$y = \dfrac{-1}{2}$	$a = 1$	$x = 8$

4 In each part, form an equation to represent the given information and solve it to find the unknown.

a I think of a number, multiply it by 8 and add 5. This gives me the same answer as when I subtract the number from 50. What is the number?

b If the areas of the triangle and trapezium are equal, find x.

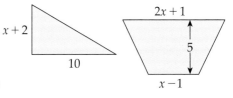

5 a Simplify $\dfrac{x + 1}{3} + \dfrac{x}{7}$ b Hence, solve $\dfrac{x + 1}{3} + \dfrac{x}{7} = \dfrac{x - 1}{4}$

Remember how to add fractions

$$\dfrac{2}{5} + \dfrac{3}{8} = \dfrac{2 \times 8 + 3 \times 5}{5 \times 8}$$

6 This equation has two solutions. Can you find them both?

$$\dfrac{25}{x + 2} = x + 2$$

Can you write an equation of your own with no solutions or with two solutions?

○ **Simultaneous equations** are solved at the same time to find the value of more than one **variable**.

Example

Alex plays a computer game. He scored 31 points by destroying two planets and three satellites and then 45 points by destroying two planets and five satellites. How many points are scored for destroying a planet and for a satellite?

You have two unknowns: the number of points scored for destroying a planet and the number of points scored for destroying a satellite.

Let x = the number of points for a planet

y = number of points for a satellite.

$2x + 3y = 31$ (1)

$2x + 5y = 45$ (2)   Label the equations.

$0 + 2y = 14$ Subtract (1) from (2) to eliminate x.

$\Rightarrow \qquad y = 7$

$2x + 21 = 31$ Substitute $y = 7$ into (1).

$\Rightarrow \qquad 2x = 10$

$x = 5$

5 points are scored for destroying a planet and 7 points for a satellite.

Check using equation (2).

$2 \times 5 + 5 \times 7 = 10 + 35 = 45$ ✓

○ You can solve simultaneous equations using **elimination**.

Add or subtract the equations to eliminate one variable.
You must look at the the signs. If they are opposite then add, if they are the same then subtract.

Example

Solve the following pairs of simultaneous equations.

$3x - 4y = 22$ (1)

$x + 4y = 18$ (2)

When you solve equations, be extra careful with + and − signs.

Eliminate the y-terms because they have matching **coefficients**.

(1) + (2) $4x + 0 = 40$ They have opposite signs so add.

$\Rightarrow \qquad x = 10$

$10 + 4y = 18$ Substitute $x = 10$ into (2).

$\Rightarrow \qquad 4y = 8$ and $y = 2$

$3 \times 10 - 4 \times 2 = 30 - 8$ Check using (1).

$= 22$ ✓

Exercise 10b

1 Solve the following pairs of simultaneous equations by subtracting.

a $a + 6b = 31$
  $a + 3b = 16$

b $2x + y = 10$
  $5x + y = 22$

c $4x + 5y = 9$
  $4x - 2y = 2$

d $x + 2y = 4$
  $5x + 2y = 16$

e $2x + 3y = 2$
  $2x + 9y = 4$

f $5x + 4y = 47$
  $5x - 4y = 23$

Remember Same Signs Subtract (SSS).

2 Solve the following pairs of simultaneous equations by adding.

a $3x + 2y = 19$
  $8x - 2y = 58$

b $5x + 2y = 16$
  $3x - 2y = 8$

c $2a + 3b = 3$
  $7a - 3b = 24$

d $2m + 3n = 19$
  $-2m + n = 1$

e $6x + 5y = 48$
  $2x - 5y = -24$

f $7x + 5y = 6$
  $-7x + y = 18$

Don't forget to check both solutions.

3 Solve the following pairs of simultaneous equations.

a $x + y = 3$
  $3x - y = 17$

b $x + y = 9$
  $x + 2y = 16$

c $2x + 3y = 5$
  $2x - 2y = -10$

d $a + 3b = 14$
  $-a + 3b = 10$

e $8x + 5y = -34$
  $-8x + 7y = 10$

f $8x + 6y = 6$
  $12x + 6y = 12$

Problem solving

4 Leah used her mobile phone to send 6 text messages and 8 picture messages and was charged £1.92. When she sent 2 text messages and 8 picture messages she was charged £1.44. Find the cost of each type of message.

5 Use these equations to make as many pairs of simultaneous equations as possible. Solve each pair that you make.

$5x + 2y = 11$ $x + 2y = 3$ $x - 2y = -5$

6 The difference between two numbers is 9 and their sum is 25. Find the two numbers.

7 Alex plays a computer game where he destroys planets and stars. Destroying two planets and five stars and gaining 30 points could be represented with the equation

  $2x + 5y = 30$.

Explain how Alex's computer game may have led to each of the following equations.

  $7x + 9y = 99$ $5x + 2y = -10$

Did you know?

The first ever text was sent by Neil Papworth on 3 December 1992. He used his computer to send the message "Merry Christmas" to a mobile phone.

Imogen bought a chocolate bar and two cereal bars for 50 pence.

Ava bought three chocolate bars and two cereal bars for 90 pence.

Since Ava spent 40 pence more and got two extra chocolate bars, each chocolate bar must cost 20 pence.

Using algebra

$$x + 2y = 50 \quad (1)$$
$$3x + 2y = 90 \quad (2)$$
$$(2) - (1) \quad 2x + 0 = 40$$
$$x = 20$$

Dev bought a chocolate bar and two cereal bars for 50 pence.

Luke bought two chocolate bars and three cereal bars for 85 pence.

×2

It helps to imagine Dev buying twice as much as he did.

$x + 2y = 50$ (1)	× 2	$2x + 4y = 100$ (3)
$2x + 3y = 85$ (2)		$2x + 3y = 85$ (2)
	$(3) - (2)$	$y = 15$

The cereal bar costs 15 pence so the chocolate bar must cost 20 pence.

> This problem is harder because the boys did not buy the same quantity of one of the items.

● To eliminate a variable you may have to multiply one or both simultaneous equations to make its coefficient the same in both.

Example

Solve the following pairs of simultaneous equations

a $x + 3y = 16$ (1)
  $2x - y = -3$ (2)

b $5x - 3y = 13$ (1)
  $3x + 2y = 4$ (2)

> Alternatively for **b** you could do 5 × (2) – 3 × (1) and eliminate x.

a Eliminate the x-terms.

  $2 \times (1) \quad 2x + 6y = 32 \quad (3)$

  All terms have been doubled.

  $2x - y = -3 \quad (2)$
  $(3) - (2) \quad 7y = 35$
  $\Rightarrow \quad y = 5$
  and $x = 1$

b Eliminate the y-terms.

  $2 \times (1) \quad 10x - 6y = 26 \quad (3)$

  $3 \times (2) \quad 9x + 6y = 12 \quad (4)$
  $(3) + (4) \quad 19x = 38$
  $\Rightarrow \quad x = 2$
  and $y = -1$

Exercise 10c

1 Solve the following pairs of simultaneous equations by first
   multiplying one of the equations.

   a $x + y = 5$ **b** $2x + y = 7$ **c** $2x - y = 9$ **d** $5a - b = 12$

       $2x + 3y = 14$ $4x + 3y = 20$ $3x + 2y = 17$ $7a - 3b = 20$

2 Solve the following pairs of simultaneous equations by first
   multiplying both of the equations.

   a $2x + 3y = 12$ **b** $2x - 3y = 19$ **c** $5a - 2b = 11$ **d** $6m - 3n = 0$

       $3x + 2y = 13$ $7x + 4y = 23$ $3a + 3b = -6$ $5m + 4n = 13$

3 Use simultaneous equations to work out the value of each
   symbol in this puzzle.

 + + + + = 69

 + + + + (smiley) = 66

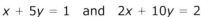

Problem solving

4 Write a pair of simultaneous equations with solutions

   a $x = 3$ and $y = 4$.

   b $m = -3$ and $n = \frac{1}{2}$.

5 Simran has two folders on her computer. One contains three songs
   and two photos and is 570 KB in size. The second contains four
   songs and five photos and is 900 KB in size. Assuming each song
   and each photo are equal in size, how big is each type of file?

6 Write a problem that could be represented by the
   simultaneous equations

       $2x + 5y = 10$ and $7x + 6y = 58$.

   Solve these equations. Does the solution make sense for the
   problem you have written? Explain.

7 Try solving these simultaneous equations and explain what
   goes wrong and why this happens. How many solutions are possible?   **‹ p.92**

       $x + 5y = 1$ and $2x + 10y = 2$

8 Use graphs to explain why these simultaneous equations have no solution.

       $y = 2x + 4$ $y = 2x + 6$

Did you know?

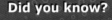

Companies
use systems of
simultaneous
equations to work
out how to make the
most profit or cause
the least pollution.

> Simultaneous equations can be **constructed** in order to solve problems in which there are two unknowns.

Example

One week Claire earns £6.78 for delivering 18 magazines and 43 newspapers.
The next week she earns the same amount for delivering 6 magazines and 52 newspapers.
How much does she earn for delivering each item?

Let x = the money earned for delivering a magazine.
y = the money earned for delivering a newspaper.

$$18x + 43y = 678 \quad (1)$$
$$6x + 52y = 678 \quad (2)$$

$3 \times (2)$ $18x + 156y = 2034 \quad (3)$ Multiply equation (2) by 3.
$$18x + 43y = 678 \quad (1)$$

$(3) - (1)$ $0 + 113y = 1356$ Subtract equation (1) from
$$y = 12$$ equation (3).

Substitute $y = 12$ into (1).
$$18x + 516 = 678$$
$$18x = 162$$
$$x = 9$$

Check the solution using (2).
$$LHS = 6 \times 9 + 52 \times 12$$
$$= 54 + 624$$
$$= 678 = RHS \checkmark$$

Interpret the solution in terms of the problem. Don't just quote $x = 9$ and $y = 12$.

Claire earns 9p for each magazine and 12p for each newspaper delivered.

Example

The perimeter of this triangle is 70 cm.
How long is each side?

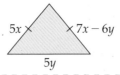

It may not be obvious that simultaneous equations are needed. Here you also need to use properties of isosceles triangles.

Isosceles triangle
$$5x = 7x - 6y \qquad \Rightarrow \qquad 2x - 6y = 0 \quad (1)$$
Perimeter = 70
$$5x + 5y + 7x - 6y = 70 \qquad \Rightarrow \qquad 12x - y = 70 \quad (2)$$
$6 \times (2)$ $72x - 6y = 420 \quad (3)$
$(3) - (1)$ $70x = 420$
$$x = 6, y = 2$$
$5y = 10$; $5x = 30$; $7x - 6y = 42 - 12 = 30$ so sides are 10 cm, 30 cm and 30 cm

Exercise 10d

1 Match the problem with the pair of simultaneous equations
that represent it and with their solution.
Interpret each solution in terms of the problem given.

A Two coffees and two teas cost £1. Four coffees and a tea cost £1.40.	**a** $2x + 2y = 100$ $x - y = 20$	**α** $x = 35, y = 15$
B One angle is 20 degrees more than another and their sum is 100.	**b** $2x + 2y = 100$ $4x + 2y = 180$	**β** $x = 60, y = 40$
C Coaching for four gym and two pool sessions costs £100. Four gym sessions cost £20 more than two pool sessions.	**c** $x + y = 100$ $x - y = 20$	**γ** $x = 30, y = 20$
D The sum of double two numbers is 100. Their difference is 20.	**d** $2x + 2y = 100$ $4x + y = 140$	**δ** $x = 15, y = 20$
E Hitting two planets and two satellites in a computer game scores 100 points. Hitting four planets and a two satellites scores 180 points.	**e** $4x + 2y = 100$ $4x - 2y = 20$	**ε** $x = 40, y = 10$

Problem solving

2 Find the solution to each problem.
   a The sum of two numbers is 34 and their difference is 6.
     What is their product?
   b Grace told her mum that the sum of their ages is half
     a century and the difference between them is two dozen.
     Is Grace a teenager?
   c A theatre takes £700 in ticket sales for a show where
     80 children and 30 adults are in the audience and £950 for a
     show where 30 children and 80 adults are in the audience.
     What is the price of a child's ticket?
   d The perimeter of this square is 80 cm.
     Find x and y.

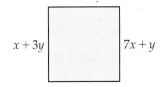

3 a Invent a problem that could be represented by
       $x + 4y = 16$
       $3x + y = 16.$
   b Solve and interpret your solution.

4 The line $y = mx + c$ passes through the points (1, 4) and (3, 14).
   Find the equation of the line.

You can find the solution to a pair of simultaneous equations by drawing a graph.

Example

Draw the graphs $2x - y = 5$ and $x + 2y = 5$ and find their point of **intersection**. What does this point represent?

$2x - y = 5$

$\Rightarrow y = 2x - 5$

x	0	1	2	3
y	-5	-3	-1	1

$x + 2y = 5$

x	0	5
y	2.5	0

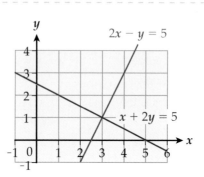

To draw a line graph plot at least two points and join the points with a ruler.

The graphs intersect at the point (3, 1).

$x = 3$ and $y = 1$ is the solution to the pair of simultaneous equations.

⬤ You can solve a pair of simultaneous equations by drawing the lines on a graph and finding their point of intersection.

Example

Use this graph to solve

$x + 3y = 8$ and $2x + y = 1$

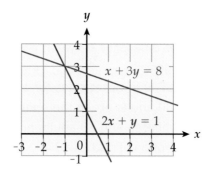

The graphs $x + 3y = 8$ and $2x + y = 1$ intersect at (-1, 3).

The solution is $x = -1$ and $y = 3$

Give the value of x and y not just the coordinates of the point of intersection.

Exercise 10e

1 Use the graph to solve these pairs of simultaneous equations.

a $y = x - 1$
$x + y = 3$

b $x + y = 3$
$y = x + 3$

c $y = x + 3$
$x + 2y = 3$

d $x + y = 3$
$x + 2y = 3$

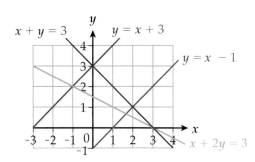

2 Plot your own graphs in order to solve these pairs of simultaneous equations.

a $y = x - 1$
$y = 2x - 5$

b $x + y = 8$
$y = x + 4$

c $y = 4x - 1$
$2x + y = 2$

d $2x + 3y = 12$
$y = x - 1$

Problem solving

3 Algie uses his mobile phone to make two calls and send five text messages. He spends 74 pence. Nicholas spends 68 pence on four calls and two text messages. They are both on the same phone tariff.

a Write two equations to represent the given information.

b Using a graphical method, find the cost of a call and of a text message.

c Check your solution using an algebraic method.

4 The lines in the diagram intersect at the point (3, 5). Write simultaneous equations which have $x = 3$ and $y = 5$ as the solution.

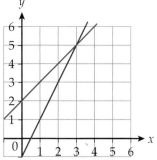

Did you know?

Architects and engineers can use straight lines to create curved surfaces. The Kobe Port Tower in Japan uses a surface called a hyperboloid.

5 Using a graph to help you, explain why the simultaneous equations $y = 3x + 1$ and $y = 3x + 5$ have no solution.
Can you give an example of a pair of simultaneous equations with two solutions?
Is it ever be possible to have more than two solutions?

6 Give a rule to say if the following pairs of equations have no, one or many solutions.

a $y = m_1x + c_1$
$y = m_2x + c_2$

b $a_1x + b_2y = c_1$
$a_2x + b_2y = c_2$

Inequalities are mathematical statements containing one or more of the symbols $<$, $\leq$, $>$ or $\geq$.

< less than
≤ less than or equal to
> greater than
≥ greater than or equal to

Since an inequality can have more than one solution, the **solution set** can be represented on a number line.

| Under 3s travel for free | Bridge not suitable for lorries over 3 m | You must be over 1.5 m to ride this roller coaster | You must be at least 15 years old to enter **Y Factor** |

  $A < 3$
0 1 2 3 4

$h \leq 3$
0 1 2 3 4

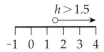 $h > 1.5$
-1 0 1 2 3 4

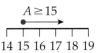 $A \geq 15$
14 15 16 17 18 19

The solid circle shows that the end number is part of the solution set and the hollow circle shows that it is not.

You can solve an inequality in a similar way to solving an equation since most operations can be applied to each side.

However if you multiply or divide an inequality by a negative number the inequality sign must be reversed.

$3 > 10$ Multiply both sides by -2 gives $-6 < -20$ FALSE
So reverse the inequality sign $-6 > -20$ TRUE

Example

a Solve these inequalities and represent the solutions on a number line.
   i $5x + 5 > 2x + 1$ **ii** $-5y \geq -20$

b Find the **integers** that satisfy $10 < 2x \leq 17$

You can think of $10 < 2x \leq 17$ as two separate inequalities $2x > 10$ and $2x \leq 17$.

a **i** $5x + 5 > 2x - 1$ $- 2x$
$3x + 5 > -1$ $- 5$
$3x > -6$ $\div 3$
$x > -2$

-3 -2 -1 0 1 2

ii $-5y \geq -20$ $\div (-5)$ and reverse the inequality.
$y \leq 4$

0 1 2 3 4 5

b $10 < 2x \leq 17$ $\div 2$
$5 < x \leq 8\frac{1}{2}$
The integers are 6, 7 and 8.

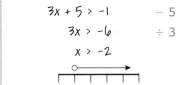
Integers are the numbers ..., -3, -2, -1, 0, 1, 2, ...

Exercise 10f

1 Decide, with a reason, if each of these inequalities is true or false.

   a $3^2 > 2^3$ **b** 10% of 450 $\geq$ 20% of 225   **c** $-\dfrac{1}{3} > -\dfrac{1}{2}$

   d $\sqrt{25} \leq 30 < \sqrt[3]{8000}$ **e** $(-5)^3 \geq 5^3$ **f** $\sqrt[4]{81} < \sqrt[3]{-64}$

2 Match the three number lines to one of the given inequalities.
For the remaining inequalities, draw their number lines.

A   -5 -4 -3 -2 -1 0   **B**   1 2 3 4 5 6   **C**   1 2 3 4 5 6

   a $x \geq 3$ **b** $x > 3$ **c** $x < 2$ **d** $x \leq 0$

   e $x > -4$ **f** $x \leq -2$ **g** $2 < x \leq 5$ **h** $2 \leq x < 5$

3 Solve these inequalities and illustrate your solutions on a number line.

   a $3x - 5 < 16$ **b** $5y + 2 \leq 3y + 8$ **c** $4z - 3 \geq 3(z - 5)$ **d** $6 > 3x$

   e $-5a \leq 30$ **f** $10 - 2x < 3$ **g** $\dfrac{x + 7}{2} \leq \dfrac{x - 1}{5}$ **h** $36 \leq 6m < 72$

4 Use inequalities to write these real-life statements.

   a (40) speed limit   **b** Have you got talent? Contestants over 16 needed   **c** Dresses to fit ladies 155 to 175cm   **d** Free Entry for children under 3 and for over 60s

   Use *S* for speed.   Use *A* for age.   Use *h* for height.   Use *A* for age.

Problem solving

5 **a** Find all the integers that satisfy $2x \geq 10$ and $3x - 2 < 31$.

    b Explain why there are no integers such that $y \geq 10$ and $2y < 8$.

6 The number of square centimetres in the area of this
rectangle is greater than the number of centimetres in
its perimeter. Write an inequality to represent this
information and solve it to find the range of values of *x*.

    $x - 1$ 6

7 If the mean of the first set of expressions is less
than the mean of the second set, write and solve
an inequality to find the possible values for *x*.

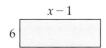

    $2x$ 10 $3x$ 4 x

    $7x$ 2 $-x$ 5 $8x$

8 Investigate the $\boxed{x!}$ button on your calculator.
Can you work out what it does?
Find the smallest value of *x* such that $x! > 1\,000\,000$

Linear equations can be solved using algebraic methods, but with more difficult equations you sometimes need to use **trial-and-improvement** to find a **solution**.

Computers use trial-and-improvement to solve complicated equations.

Example

Use trial-and-improvement to solve $x^2 + x = 48$, giving your answer to one decimal place.

	A	B	C	D
	x	x^2	$x^2 + x$	Result
1				
2	6	36	42	low
3	7	49	58	high
4	6.4	40.96	47.36	low
5	6.5	42.25	48.75	high
6	6.45	41.6025	48.0525	high

$6.4 < x < 6.45$

$x = 6.4$ (1dp)

The answer is 6.4 because the answer lies between 6.40 and 6.45, so it rounds to 6.4 (1dp).

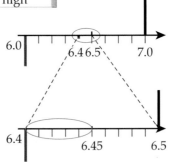

Organise trials to sandwich the solution. Always work to one more decimal place than asked for.

You may need to first construct an equation, perhaps using formulae from other areas of mathematics, then solve it using trial-and-improvement.

Example

The volume of this cuboid is 392 cm³.
How long is each side?

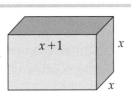

Volume of cuboid = lwh

So, $x \times x \times (x + 1) = 392$ $\Rightarrow$ $x^2(x + 1) = 392$

x	x^2	$x + 1$	$x^2(x + 1)$	Result
5	25	6	150	low
6	36	7	252	low
7	49	8	392	exact

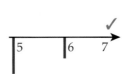

$x = 7$

The width and height are both 7 cm and the length is 8 cm.

Remember BIDMAS.
$x \times x \times x + 1$ is not the same as
$x \times x \times (x + 1)$

Exercise 10g

1 Select four equations that you might need trial-and-improvement to solve.

a $x^2 - 2x = 40$ **b** $5x + 4 = 2 - x$ **c** $x^3 + x = 68$

d $x(x + 2) = 10$ **e** $x^2 = 25$ **f** $\sqrt{x} - \dfrac{1}{x} = 5$

2 Find the exact solutions of the following equations using trial-and-improvement.

a $x^2 - 2x = 255$ **b** $3x^3 + x = 1544$ **c** $y(y + 5) = 644$

d $2^x = 33\,554\,432$ **e** $x^3 + (x - 1)^3 = 1729$ **f** $2^x = 16(3 - x)^2$

3 Use trial-and-improvement to solve these equations to 1 dp.

a $x^2 + 4x = 100$ **b** $x^3 - x = 700$ **c** $5x^2 + 3x = 90$

d $y(y - 2) = 220$ **e** $x^2 - x - 1 = 0$ **f** $2^x - 1000 = 0$

4 Use trial-and-improvement to solve these equations to the given number of decimal places.

a $3x^2 - x = 280$ (2 dp) **b** $3^x = 42$ (2 dp) **c** $\sqrt{x} + x = 90$ (3 dp)

d $x^4 + 2x = 630$ (3 dp) **e** $5x^3 - 30x^2 - 1 = 0$ (2 dp) **f** $x^3 = 2$ (4 dp)

Problem solving

5 Construct an equation and solve it, using trial-and-improvement, to find the exact solution in each case.

a The product of three consecutive numbers is 32 736.

b The sum of the square of a number and its square root is 4104.

6 Two students used trial-and-improvement to solve $x^2 + 3x = 60$ to 1 dp. List all of the mistakes that each student has made.

x	x^2	$3x$	$x^2 + 3x$
6	36	18	54
6.8	46.24	20.4	66.64
6.2	38.44	18.6	57.04
6.4	40.96	19.2	60.16

$x = 60.16$

x	x^2	$3x$	$x^2 + 3x$
6	36	18	54
7	49	21	70
6.3	39.69	18.9	58.59
6.4	40.96	19.2	60.16
6.35	40.3225	19.05	59.3725
6.36	40.4496	19.08	59.5296
6.37	40.5769	19.11	59.6869
6.38	40.7044	19.14	59.8444
6.39	40.8321	19.17	60.0021

$x = 6.4$

7 The volume of this cuboid is 780 cm³. Find x.

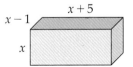

8 Many equations on this page are quadratic equations. What is a quadratic equation and do you really need trial-and-improvement to solve them?

10 MySummary

Check out

You should now be able to ...

Test it ➡

Questions

✓ Solve linear equations with brackets and algebraic fractions.	7	1
✓ Solve simultaneous equations by elimination.	7	2 – 5
✓ Solve simultaneous equations by drawing graphs.	8	6
✓ Solve linear inequalities with one variable.	7	7, 8
✓ Find approximate solutions to equations using trial-and-improvement.	6	9

Language	Meaning	Example
Simultaneous equations	A set of equations that have a common solution.	$y = 3x - 8$ $y = 2x - 4$ has solutions $x = 4$, $y = 4$
Elimination	A method to solve simultaneous equations by addition or subtraction.	$y = 3x - 8 \Rightarrow 0 = x - 4$ $y = 2x - 4$ so $x = 4$
Intersection	A point at which two lines cross.	$y = 3x - 8$ and $y = 2x - 4$ intersect at the point (4, 4)
Inequality	A relationship between two unequal quantities.	$4 < 7$
Solution set	The set of solutions to an inequality or equation.	$3 < y \leq 7$ has integer solution set {4, 5, 6, 7}

1 Solve these equations.

  a $3a - 40 = -1$

  b $2(3b + 9) = 3(27 + b)$

  c $5 - 2c = 11 - 6c$

  d $\dfrac{d + 7}{3} + 4 = -\dfrac{5d}{4}$

  e $\dfrac{5e + 7}{5} = \dfrac{3 - 10e}{7}$

  f $\dfrac{3}{7f - 1} = \dfrac{5}{13 - 14f}$

2 Solve these pairs of simultaneous equations.

  a $x + 3y = 23$
    $x - 2y = -12$

  b $5x + 2y = 7$
    $7x + 2y = 5$

  c $2x - 4y = -5$
    $8x - 4y = -2$

  d $x - 12y = -10$
    $x + 12y = 2$

3 Solve these pairs of simultaneous equations.

  a $2x + y = 8$ **b** $3x + 5y = 23$
    $x - 3y = 11$ $x + 7y = 45$

  c $x - y = 3$ **d** $13x - 2y = 53$
    $4x - 7y = 39$ $12x + 4y = 46$

4 Solve these pairs of simultaneous equations.

  a $2x + 3y = 21$ **b** $5x + 5y = 20$
    $3x - 2y = 12$ $2x + 7y = 38$

  c $3x - 7y = 5$ **d** $15x - 13y = 0$
    $4x - 3y = 3.5$ $10x + 39y = 22$

5 Sam's mum was 25 when he was born and the sum of their ages now is 53 years. How old is Sam now?

6 Plot the graphs and use to solve the pairs of simultaneous equations.

  a $y = 2x - 3$ **b** $5x + 3y = 15$
    $x + y = 9$ $y = x - 1$

7 Solve these inequalities and illustrate your solutions on a number line.

  a $6(2x + 1) \le 5x - 1$

  b $15 - 2x > 3$

8 **a** Find all the integers that satisfy both $3x \le 21$ and $2x + 5 > 13$.

  b Illustrate your solution on a number line.

9 Use a trial-and-improvement method to solve these equations to 2 dp.

  a $x^3 - x = 1610$

  b $\sqrt{3x} + 2x = 5$

What next?

Score		
	0 – 4	Your knowledge of this topic is still developing. To improve look at Formative test: 3C-10; MyMaths: 1182, 1176, 1175, 1236, 1174, 1319, 1162, 1161 and 1057
	5 – 7	You are gaining a secure knowledge of this topic. To improve look at InvisiPen: 232, 236, 237, 241, 243, 244 and 245
	8 – 9	You have mastered this topic. Well done, you are ready to progress!

1 Solve the following equations.

    a $5a + 11 = 3a - 2$ **b** $10 - 5x = 2(3 - 2x)$

    c $15 - 2y = 25$ **d** $\dfrac{5}{2x - 1} = \dfrac{2}{x + 3}$

2 The means of the numbers in each box are equal.
By first using an equation to find x, find this mean.

10	x	-6
$5x$	2	$4x$

7	$4x$	-1
$-12x$	15	

3 Solve the following pairs of simultaneous equations.

    a $5x + 2y = 6$ **b** $3a + b = 14$ **c** $5m + 3n = 16$

         $7x + 2y = 8$ $3a + 6b = 9$ $3m - 3n = 0$

    d $4x - y = 13$ **e** $19x + 17y = 14$ **f** $8x + 5y = 3$

         $8x - y = 21$ $-3x - 17y = 140$ $12x + 5y = 4$

4 Solve the following pairs of simultaneous equations.

    a $5x + y = 7$ **b** $2a + 7b = -15$ **c** $5m + 3n = 17$

         $3x + 2y = 7$ $4a - 5b = 27$ $3m - 5n = -17$

    d $3x - 8y = 10$ **e** $7p + 4q = 30$ **f** $6s - 12t = 1$

         $2x - 6y = 7$ $5p - 3q = 39$ $-12s + 6t = 4$

5 Construct and solve a pair of simultaneous equations to find the solution to each problem.

    a   Two compact discs and four DVDs cost £60. Five compact discs and three DVDs cost £66. How much does a DVD cost?

    b   The perimeter of this rectangle is 22 cm. What are its dimensions?

$x - 2y$ $2x + 7y$

         8

    c   At half-term two groups of students and teachers set off on school trips. 71 people on the Year 9 trip just fill three minibuses and five cars, whilst 33 people on the sixth form trip fill one minibus and four cars. How many people can each type of vehicle hold?

6 The graph shows the lines
$y = 2x + 3$ and $x + 2y = 1$.
Use the graph to solve
$y = 2x + 3$ and $x + 2y = 1$ simultaneously.

7 Solve graphically these simultaneous equations.
  a $y = 5x - 2$ and $y - 2x = 4$
  b $3x + 2y + 6 = 0$ and $x = 2 - 2y$

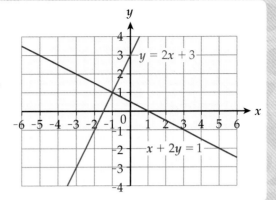

8 For each inequality
  i draw a number line to represent the inequality
  ii state at least three integers that satisfy the inequality.
  a $x > 3$ **b** $y \leq -2$ **c** $m < 8$
  d $5 < x \leq 8$ **e** $7 > p$ **f** $-2 \leq x < 1$
  g $-x \geq 7$ **h** $6 \geq x > -1$ **i** $x > 4$ and $x \geq -2$

9 Solve the following inequalities and show each solution
set on a number line.
  a $4x + 8 > 2x + 20$ **b** $6(2 - 3y) \leq 3(3 - 3y)$ **c** $-5x < 20$
  d $10 - 7x < 3$ **e** $3(2x + 1) > 4(7 - x)$ **f** $-8x - 12 \leq -20$

10 The product of two consecutive integers is 1056.
  a Write an equation to represent this information.
  b Solve this equation using trial-and-improvement to find the
     consecutive integers.

11 Solve the following equations, to the required number of decimal places,
using trial-and-improvement.
  a $x^2 + x = 80$ (1 dp)
  b $x(x + 2) = 194$ (2 dp)

12 A rectangle's length exceeds its width by 3 cm.
  a Write an expression for the length of the rectangle if the width is x.
  b If the area of the rectangle is 100 cm^2, form an equation in x.
  c Use trial-and-improvement to find the dimensions of the rectangle to
     1 decimal place.

11 Powers and roots

Introduction

When you look up at the night sky you might see the M31 Galaxy, more commonly known as the Andromeda Galaxy. It is about 24 000 000 000 000 000 000 km from the Earth. Astronomers write this distance in **standard index form** as 2.4×10^{19} km. The light from the M31 galaxy has taken about 2.3 million years to reach the earth, so you are effectively looking back in time!

What's the point?

Scientists and astronomers have to work with very large and very small numbers. Standard index form was invented so that a number like the distance to the M31 galaxy can be written down in a much easier and clearer way than writing lots of zeros.

Objectives

By the end of this chapter, you will have learned how to …
● Write numbers in standard form.
● Calculate with standard form.
● Know and use the index laws.
● Know and use rules for surds.
● Use index notation for square and cube roots.

Check in

1 Calculate

   a 3.8×0.01 **b** $2.75 \div 0.1$ **c** 0.00045×100 **d** $0.001 \div 0.00001$

2 Round each of these numbers to two significant figures.

   a   17.475   **b**   0.7468   **c**   2010   **d**   0.04026

Starter problem

Distance of Earth to Moon: 384 000 km

Length of 1 stride: 80 cm

How many steps would you need to take to walk to the moon?

How long would it take you?

What if you had to walk to the Sun, the nearest star, the nearest galaxy...?

Investigate.

● The decimal system is based upon **powers** of 10. You can write all powers of ten using **index notation**.

The power is also called an index.

1 million (mega)	= 1 000 000	= 10 × 10 × 10 × 10 × 10 × 10 = 10^6	
1 thousand (kilo)	= 1000	= 10 × 10 × 10	= 10^3
1 hundred	= 100	= 10 × 10	= 10^2
1 ten	= 10	= 10	= 10^1
1 unit	= 1		= 10^0
1 tenth	= $\frac{1}{10}$ = 0.1	= $\frac{1}{10^1}$	= 10^{-1}
1 hundredth (centi)	= $\frac{1}{100}$ = 0.01	= $\frac{1}{10^2}$	= 10^{-2}
1 thousandth (milli)	= $\frac{1}{1000}$ = 0.001	= $\frac{1}{10^3}$	= 10^{-3}

● To multiply by a positive power of ten, 10^n, move the digits n places to the left of the decimal point.

Any number to the power of zero is equal to 1.

Example

Write out in full

a 3.3×10^3 **b** 2.75×10^5

3.3×10^3 and 2.75×10^5 are in **standard index form**.

a $3.3 \times 10^3 = 3.3 \times 1000$

           = 3300

b $2.75 \times 10^5 = 2.75 \times 100\,000$

           = 275\,000

The digits move to the left.

You use powers of 10 to write very large numbers.

● To write a large number in standard index form, re-write it as a number between 1 and 10 multiplied by a positive power of 10.

Example

Write these numbers in standard index form.

a   6700   **b**   26 000 000

Divide the number by a power of 10 so that it becomes a decimal between 1 and 10.

a $6700 \div 1000 = 6.7$ **b** $26\,000\,000 \div 10\,000\,000 = 2.6$

      $6700 = 6.7 \times 1000$ $26\,000\,000 = 2.6 \times 10\,000\,000$

           $= 6.7 \times 10^3$ $= 2.6 \times 10^7$

The digits move to the right.

Exercise 11a

1 Each of these numbers is in standard index form. Write out each number in full.

   a 3.7×10^2 **b** 4.7×10^3

   c 1.23×10^6 **d** 4.02×10^7

   e 3.01×10^6 **f** 4.9×10^9

   g 7.37×10^{11} **h** 1.004×10^6

2 Write each of these numbers in standard index form.

   a 230   **b** 4870

   c 340 000   **d** 78 000 000

   e 4 100 000 000   **f** 2 380 000

   g 238.3   **h** 3878.8

3 Calculate the following, giving your answers in standard form.

   a $8.6 \times 10^4 + 2.2 \times 10^4$

   b $2.4 \times 10^3 + 85$

> Write out the numbers in full before adding or subtracting.

   c $3.8 \times 10^4 - 3.3 \times 10^3$

   d $7.6 \times 10^5 - 3.3 \times 10^3$

4 Write each of these numbers in standard index form.

   a 27×10^1 **b** 573×10^2

   c 0.53×10^4 **d** 34.2×10^3

   e 3010×10^4 **f** 0.492×10^6

   g 0.048×10^5 **h** 0.000378×10^8

5 Round each of these numbers to the nearest thousand. Give your answers in standard form.

   a 5.3×10^3 **b** 2.85×10^4

   c 4.1365×10^5 **d** 3.5072×10^4

Problem solving

6 Write each of these numbers in standard index form.

   a The population of Maryport is 35 600.

   b The population of the UK in 1990 was 59 million.

   c The speed of light is 186 000 miles per second.

   d The nearest star is 24 790 000 000 000 miles away.

7 Put these numbers in order from smallest to largest.

   a 2.6×10^3, 2.5×10^4, 270, 2.55×10^4, 2.58×10^3

   b 3×10^{12}, 2.8×10^{14}, 2.9×10^{13}, 2 980 000 000 000 000

8 **a** Carla flies 3.6×10^3 miles in 6 hours. What is the average speed of the plane?

   b Keiley wins €1.3×10^7 in a lottery. She spends €3.85×10^5 on a new house and €32 500 on a new car. How much money does she have left?

9 Complete these statements. The first one is done for you.

   a $5 \times 10^n = 50 \times 10^{\square} = 50 \times 10^{n-1}$

   b $0.5 \times 10^n = 5000 \times 10^{\square}$

   c $500 \times 10^n = 5 \times 10^{\square}$

   d $500 000 \times 10^n = 50 \times 10^{\square}$

‹ p.34

● To multiply (or divide) by a negative power of ten, 10^{-n}, move the digits n places to the right (or left) of the decimal point.

Example

Calculate

a 330×10^{-3} **b** $27.5 \div 10^{-2}$.

a $330 \times 10^{-3} = 330 \times \dfrac{1}{1000}$
               $= 330 \div 1000$
               $= 0.33$

b $27.5 \div 10^{-2} = 27.5 \div \dfrac{1}{100}$
               $= 27.5 \times 100$
               $= 2750$

In index notation 10^{-3} is the same as $\frac{1}{10^3} = \frac{1}{1000}$ and multiplying by $\frac{1}{1000}$ is the same as dividing by 1000.

● To write a small number in standard index form, write it as a number between 1 and 10 multiplied by a negative power of 10.

Example

Write these numbers in standard index form.

a   0.67   **b**   0.000026

Rewrite the number as a decimal between 1 and 10 divided by a power of 10.

a $0.67 = 6.7 \div 10$
          $= 6.7 \div 10^1$

b $0.000026 = 2.6 \div 100\,000$
                 $= 2.6 \div 10^5$

Write the number in standard form using a negative power of 10.

         $= 6.7 \times \dfrac{1}{10^1} = 6.7 \times 10^{-1}$
                      $= 2.6 \times \dfrac{1}{10^5} = 2.6 \times 10^{-5}$

Dividing by 10^5 is the same as multiplying by $\frac{1}{10^5} = 10^{-5}$

● You can use a calculator to work with numbers written in standard form by using the [exp] or [×10▫] button.

Example

Calculate $(3.2 \times 10^4) \times (2.5 \times 10^{-2})$.

[3] [.] [2] [exp] [4] [×] [2] [.] [5] [exp] [(−)] [2] [=]

```
3.2ᴱ4 x 2.5ᴱ-2
        800.
```

In standard form the answer is written.

8×10^2

Your calculator might have a different key for power of 10. Make sure you know how to use it.

Exercise 11b

1 Calculate the following.

 a $47 \div 10^2$ **b** $2900 \div 10^3$
 c $123 \div 10^4$ **d** $4 \div 10^5$
 e $31.8 \div 10^{-4}$ **f** $39\,000 \times 10^{-2}$
 g 0.3×10^{-3} **h** 2.4×10^{-2}

2 Each of these numbers is in standard index form.
Write each number as a decimal.

 a 2.8×10^{-2} **b** 3.6×10^{-3}
 c 9.34×10^{-4} **d** 5.13×10^{-3}
 e 4.92×10^{-5} **f** 3.8×10^{-6}
 g 6.25×10^{-8} **h** 1.234×10^{-10}

3 Write each number in standard index form.

 a 23×10^2 **b** 112×10^{-4}
 c 0.2×10^7 **d** 0.067×10^{-5}

4 Write each number in standard index form.

 a 0.3   **b** 0.48
 c 0.034   **d** 0.00078
 e 0.000003   **f** 0.0067
 g 0.00000456   **h** 0.000000000024

5 Put these numbers in order from smallest to largest.

 2×10^{-8} 1.8×10^{-7}
 2.3×10^{-8} 2.2×10^{-7}

6 Use your calculator to evaluate the following, giving your answers in standard form.

 a $(6.5 \times 10^3) \times (2.4 \times 10^5)$
 b $(6.5 \times 10^3) + (2.4 \times 10^5)$
 c $(6.5 \times 10^{-3}) \div (2.4 \times 10^{-5})$
 d $(6.5 \times 10^{-3}) - (2.4 \times 10^{-5})$

Problem solving

7 Write each of these numbers in standard index form.

 a A millipede is 0.094 inches long.
 b The diameter of a human blood cell is 0.00000068 m.

8 Calculate the following, giving your answers in standard form.

 a A light year is 9.43×10^{12} km. A star is 12.4 light years from Earth. How far away is the star in km?
 b The population of India is 1.05×10^9 and India has a land area of 3.3×10^6 km².
 On average how many people live in each square kilometre in India?

9 Here are the probabilities of various events.

 i Winning the national lottery in a week   1 in 14 000 000
 ii Losing your bag at an airport in a year   1 in 138
 iii Being struck by lightning in a year   1 in 576 000
 iv Being bitten by a shark in a year   1 in 300 000 000
  a Write each probability in standard form.
  b Put the events in order of likelihood.
  c There are 60 million people in Britain. How many people will each of these events affect each year?

Did you know?

The most densely populated country on Earth is Monaco with 1.7×10^4 people/km². The average population density of Earth is 45 people/km².

● You can multiply and divide numbers written in index form.

❮ p.30

Example

Calculate   **a** $10^2 \times 10^3$ **b** $10^2 \div 10^4$

a $10^2 \times 10^3 = (10 \times 10) \times (10 \times 10 \times 10) = 10^5$

                    $= 10^{2+3}$ $= 10^5$

b $10^2 \div 10^4 = \dfrac{(10 \times 10)}{(10 \times 10 \times 10 \times 10)} = \dfrac{1}{10^2} = 10^{-2}$

                    $= 10^{2-4}$ $= 10^{-2}$

> Add the indices when you multiply
> $x^a \times x^b = x^{a+b}$

> Subtract the indices when you divide
> $x^a \div x^b = x^{a-b}$

● You can multiply a number by itself when it is written in index form.

Example

Calculate   **a** $(10^3)^3$ **b** $(10^{-3})^2$

a $(10^3)^3 = 10^3 \times 10^3 \times 10^3 = 10^{3+3+3} = 10^9$

                        $= 10^{3 \times 3} = 10^9$

b $(10^{-3})^2 = 10^{-3} \times 10^{-3} = 10^{-3+-3} = 10^{-6}$

                        $= 10^{-3 \times 2} = 10^{-6}$

> Multiply the indices when you raise an index to a power $(x^a)^b = x^{ab}$

❮ p.118

● When a calculation contains more than one operation, you must do the operations in the correct order.

  ▶ This is called the **order of operations**.

Example

Calculate $\dfrac{2(3 - 3^2)^2}{3 + \sqrt{4^2 - 7}}$

$\dfrac{2(3 - 3^2)^2}{3 + \sqrt{(4^2 - 7)}} = \dfrac{2(-6)^2}{3 + \sqrt{9}}$

                $= \dfrac{2 \times 36}{3 + 3}$

                $= \dfrac{72}{6}$

                $= 12$

First, work out the contents of any brackets.

Second, work out any powers or roots.

> The order of operations is **B**rackets, **I**ndices, **D**ivide, **M**ultiply, **A**dd, **S**ubtract (BIDMAS).

7

Exercise 11c

1 Simplify each of the following, leaving your answer as a single power of the number.

a $3^2 \times 3^2$ **b** $4^3 \times 4^4$

c $5^2 \times 5^6$ **d** $3^5 \times 3^{-2}$

e $7^6 \times 7^{-3}$ **f** $3^5 \div 3^5$

g $4^3 \div 4^5$ **h** $3^5 \div 3^7$

i $2^6 \div 2^{-3}$ **j** $3^5 \div 3^{-12}$

2 Calculate the following, leaving your answer in index form where possible.

a $4^{-2} \times 4^{-3}$ **b** $4^3 + 4^2$

c $5^3 - 5^2$ **d** $4^2 \div 4^3$

e $4^3 \div 4^2$ **f** $6^{-3} \times 6^{-4}$

3 Work out the value of each of the following.

a $2^2 \times 2^2$ **b** $5^4 \times 5^{-2} \times 5^{-2}$

c $3^2 \div 3^2$ **d** $\dfrac{3^4 \times 3^2}{3^8}$

e $\dfrac{4^5 \times 4^1}{4^4}$ **f** $10^2 \div 10^7$

g $\dfrac{7^3}{7^4} \times \dfrac{7^5}{7^2}$ **h** $\dfrac{1}{8^3} \div \dfrac{1}{8^6}$

4 Simplify each of the following, leaving your answer as a single power of the number.

a $2^2 \times 2^4 \times 2$ **b** $3^5 \times 3^2 \times 3^{-3}$

c $10^2 \times 10^3 \div 10^6$ **d** $\dfrac{3^4 \times 3^5}{3^3 \times 3^2}$

e $\dfrac{2^2 \times 2 \times 2^3}{2^2 \times 2^8}$ **f** $\dfrac{10^3 \div 10^{-5}}{10^2}$

5 Calculate the following, leaving your answer in index form.

a $(4^2)^3$ **b** $(5^3)^2$

c $(2^{-2})^3$ **d** $(4^2)^{-2}$

e $(10^{-3})^{-2}$ **f** $((2^3)^{-2})^4$

6 Calculate the following, giving your answer to 2 dp where appropriate.

a $\dfrac{(5 + 3)^2}{(5 - 3)^3}$ **b** $\dfrac{(2^3 - 1)(7 + 2)^2}{9 - 4^2}$

c $\dfrac{(3^2)^2 + \sqrt{3^4}}{(10 - 7)^2}$ **d** $\dfrac{2.7 \times 11.2}{1.05^2 \times (3.1 - 2)^2}$

e $\dfrac{4 \times \sqrt{1.3^3 + 4^2}}{3.7 - 1.2^2}$ **f** $\dfrac{\sqrt{(3^2 + 2)}}{2\sqrt{(0.4^2 + 1)}}$

Problem solving

7 Give four different values for a and b which satisfy this equation.

$$a^b = 64$$

8 George knows that $p^2 = 2$. Use this information to find the value of

a p^4 **b** p^{-2} **c** p^6 **d** $(p^2)^2$.

9 Work out the following, without using a calculator, giving your answers in standard form.

a $(3 \times 10^3) \times (2 \times 10^5)$

b $(6 \times 10^5) \div (2 \times 10^3)$

c $(3 \times 10^{-3}) \times (2 \times 10^5)$

d $(6 \times 10^{-3}) \div (2 \times 10^2)$

Use your calculator to check your answers.

Write some rules for multiplying and dividing numbers in standard form without using a calculator.

Did you know?

French mathematician and philosopher Rene Descartes was the first person to use index form to write powers of numbers.

● You can multiply numbers written in **surd** form.
$$\sqrt{a} \times \sqrt{b} = \sqrt{a \times b} = \sqrt{ab}$$

> Most square roots cannot be written as exact decimals
> $$\sqrt{10} = 3.1622776...$$
> It is more accurate to leave these numbers in **surd** form.
> $\sqrt{10}$ is a surd.

Example

Calculate

a $\sqrt{5} \times \sqrt{5}$

b $\sqrt{5} \times \sqrt{2}$

a $\sqrt{5} \times \sqrt{5} = \sqrt{5 \times 5}$
$$= \sqrt{25}$$
$$= 5$$

b $\sqrt{5} \times \sqrt{2} = \sqrt{5 \times 2}$
$$= \sqrt{10}$$

● You can use the multiplication rule to simplify surds.

Example

Write these surds in their simplest form.

a $\sqrt{27}$

b $\sqrt{32}$

a $\sqrt{27} = \sqrt{9 \times 3}$
$$= \sqrt{9} \times \sqrt{3}$$
$$= 3 \times \sqrt{3}$$
$$= 3\sqrt{3}$$

b $\sqrt{32} = \sqrt{16 \times 2}$
$$= \sqrt{16} \times \sqrt{2}$$
$$= 4 \times \sqrt{2}$$
$$= 4\sqrt{2}$$

● You can multiply numbers written in simplified surd form.

Example

Calculate

a $2\sqrt{5} \times 3\sqrt{5}$

b $2\sqrt{5} \times 3\sqrt{2}$

a $2\sqrt{5} \times 3\sqrt{5} = (2 \times 3) \times (\sqrt{5} \times \sqrt{5})$
$$= 6 \times 5$$
$$= 30$$

b $2\sqrt{5} \times 3\sqrt{2} = (2 \times 3) \times \sqrt{5 \times 2}$
$$= 6 \times \sqrt{10}$$
$$= 6\sqrt{10}$$

‹ p.34

● You can use **fractional indices** to represent square roots and cube roots in **index form**.

> Square root is a power of $\frac{1}{2}$.
> Cube root is a power of $\frac{1}{3}$.

Example

Write these square roots and cube roots in index notation.

a $\sqrt{4}$ **b** $\sqrt{10}$ **c** $\sqrt[3]{8}$ **d** $\sqrt[3]{10}$

a $\sqrt{4} = 4^{\frac{1}{2}}$ **b** $\sqrt{10} = 10^{\frac{1}{2}}$ **c** $\sqrt[3]{8} = 8^{\frac{1}{3}}$ **d** $\sqrt[3]{10} = 10^{\frac{1}{3}}$

8

Exercise 11d

1 Calculate the following, leaving your answers in surd form.

   a $\sqrt{3} \times \sqrt{3}$ **b** $\sqrt{4} \times \sqrt{4}$

   c $\sqrt{5} \times \sqrt{3}$ **d** $\sqrt{6} \times \sqrt{2}$

   e $\sqrt{7} \times \sqrt{3}$ **f** $\sqrt{2} \times \sqrt{18}$

   g $\sqrt{2} \times \sqrt{32}$ **h** $\sqrt{5} \times \sqrt{20}$

   i $\sqrt{27} \times \sqrt{3}$ **j** $\sqrt{8} \times \sqrt{18}$

2 Write these numbers in their simplest form.

   a $\sqrt{12}$ **b** $\sqrt{8}$

   c $\sqrt{18}$ **d** $\sqrt{24}$

   e $\sqrt{40}$ **f** $\sqrt{48}$

   g $\sqrt{32}$ **h** $\sqrt{50}$

   i $\sqrt{72}$ **j** $\sqrt{98}$

3 Calculate the following, leaving your answers in surd form.

   a $2\sqrt{3} \times 3\sqrt{3}$ **b** $3\sqrt{2} \times \sqrt{2}$

   c $4\sqrt{5} \times 2\sqrt{3}$ **d** $3\sqrt{6} \times 2\sqrt{2}$

   e $4\sqrt{7} \times 2\sqrt{3}$ **f** $3\sqrt{2} \times 4\sqrt{5}$

   g $\sqrt{2} \times 2\sqrt{3}$ **h** $\sqrt{5} \times 3\sqrt{20}$

   i $5\sqrt{18} \times 2\sqrt{2}$ **j** $6\sqrt{7} \times 3\sqrt{1}$

   k $3\sqrt{2} \times 3\sqrt{32}$ **l** $4\sqrt{3} \times 2\sqrt{27}$

4 Write these numbers using index notation.

   a $\sqrt{8}$ **b** $\sqrt[3]{8}$ **c** $\sqrt{18}$

   d $\sqrt[3]{24}$ **e** $\sqrt[3]{40}$

5 Work out the value of each of these expressions.

   a 3^2 **b** $9^{\frac{1}{2}}$ **c** $8^{\frac{1}{3}}$

   d $16^{\frac{1}{2}}$ **e** $64^{\frac{1}{3}}$ **f** $27^{\frac{1}{3}}$

Problem solving

6 Calculate the area of this triangle.

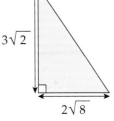

  $3\sqrt{2}$

     $2\sqrt{8}$

7 Calculate the following, leaving your answers in index form.

   a $8^{\frac{1}{3}} \times 8^{\frac{1}{2}}$ **b** $9^{\frac{1}{2}} \times 9^2$

   c $16^4 \div 16^{\frac{1}{2}}$ **d** $4^2 \div 4^{\frac{1}{3}}$

   e $4^{-3} \div 4^{\frac{1}{2}}$ **f** $16^{-\frac{1}{2}} \div 16^{\frac{1}{4}}$

> **Index laws**
> $x^n \times x^m = x^{n+m}$
> $x^n \div x^m = x^{n-m}$

8 Give four different values for a and b which satisfy this equation.

$$a^b = 4$$

9 **a** For each of the following, write the answer in index form and then state the numerical value of the answer.

    i $(4^{\frac{1}{2}})^2$ **ii** $(8^{\frac{1}{3}})^2$ **iii** $(4^{\frac{1}{2}})^3$ **iv** $(64^{\frac{1}{3}})^2$

   b Write what you have noticed.

   c Use your answer to part **b** to work out the value of $25^{\frac{3}{2}}$.

Did you know?

  $\sqrt{2}$

Around 500 BC Hippasus of Metapontum proved that $\sqrt{2}$ could not be written as a fraction. According to legend, this so upset his fellow Pythagoreans that they drowned him!

Check out

You should now be able to ...

Test it ➡

Questions

✓ Write numbers in standard form.	8	1, 2
✓ Calculate with standard form.	8	3
✓ Know and use the index laws.	7	4
✓ Know and use rules for surds.	8	5, 6
✓ Use index notation for square and cube roots.	8	7 – 9

Language	Meaning	Example
Power	A shorthand way of writing a number multiplied by itself several times.	$a \times a = a^2$ "a squared" $b \times b \times b \times b \times b = b^5$ "b to the power 5"
Index notation	Writing a number as a power.	$3 \times 3 \times 3 = 3^3$
Standard index form	A way to write very large or small numbers as a number between 1 and 10 multiplied by a power of 10.	$34\,500 = 3.45 \times 10^4$ $0.000345 = 3.45 \times 10^{-4}$
Order of operations	A set order to perform operations (BIDMAS) brackets-powers-division then multiplication-addition and subtraction.	$3 + 2^2 \times 5 = 23$
Surd	A number written as a square root.	$\sqrt{3}$

1 Write each number out in full.

  a 2.1×10^7

  b 2.69×10^5

  c 7.4×10^{-3}

  d 3.73×10^{-4}

2 Write each number in standard form.

  a $26\,000$

  b $988\,000$

  c $0.000\,088$

  d $0.007\,59$

  e 59.4×10^8

  f 0.098×10^{-7}

  g 0.14×10^5

  h 672×10^{-4}

3 Work out the following calculations without using a calculator.
Give your answers in standard form.

  a $(8.4 \times 10^6) + (9.5 \times 10^4)$

  b $(5.9 \times 10^{-2}) - (7.28 \times 10^{-3})$

4 Simplify each of the following. Leave your answer as a single power of the number.

  a $9^3 \times 9^5$ **b** $8^{-2} \times 8^{-7}$

  c $6^3 \div 6^{-4}$ **d** $3^{-9} \div 3^2$

  e $(4^7)^2$ **f** $(7^{-3})^{-5}$

  g $((3^2)^2)^3$ **h** $((4^2)^{-2})^3$

5 Calculate the following, leaving your answers in surd form where appropriate.

  a $\sqrt{7} \times \sqrt{7}$

  b $\sqrt{10} \times \sqrt{5}$

  c $\sqrt{12} \times \sqrt{18}$

  d $\sqrt{4} \times \sqrt{8}$

  e $2\sqrt{8} \times 3\sqrt{20}$

  f $6\sqrt{2} \times 2\sqrt{24}$

6 Write these numbers in their simplest form.

  a $\sqrt{32}$ **b** $\sqrt{80}$

  c $\sqrt{175}$ **d** $\sqrt{320}$

7 Write these numbers using index notation.

  a $\sqrt{7}$ **b** $\sqrt[3]{5}$

8 Work out the value of each of these expressions.

  a $49^{\frac{1}{2}}$ **b** $64^{\frac{1}{3}}$

9 Calculate the following, leaving your answers in index form where appropriate.

  a $6^{\frac{1}{2}} \times 6^{\frac{1}{2}}$ **b** $7^{\frac{1}{2}} \times 7^{-2}$

  c $5^{-\frac{1}{3}} \times 5^{\frac{1}{3}}$ **d** $8^{\frac{1}{2}} \div 8^{-3}$

  e $4^{\frac{1}{3}} \div 4^{-\frac{1}{3}}$ **f** $\left(3^{\frac{1}{2}}\right)^{-3}$

What next?

Score		
	0 – 4	Your knowledge of this topic is still developing. To improve look at Formative test: 3C-11; MyMaths: 1051, 1049, 1167, 1033, 1064 and 1065
	5 – 7	You are gaining a secure knowledge of this topic. To improve look at InvisiPen: 183, 184, 185 and 186
	8 – 9	You have mastered this topic. Well done, you are ready to progress!

1 Write these numbers in standard index form.

    a 460   **b** 813 000

    c 98 000 000   **d** 201.4

    e 75 800 000   **f** 107 500 000

    g 878 500 000 000   **h** 432 000 600

2 Write these numbers in standard index form.

    a The population of the USA is 261 million.

    b The speed of light is 297 600 kilometres per second.

    c At their most distant from each other, the Earth and Jupiter are 928 081 020 km apart.

    d Dinosaurs roamed the earth during the Jurassic Period, 199.6 to 145.5 million years ago.

    e At any given time, it is estimated there are about 10 quintillion (10 000 000 000 000 000 000) individual insects alive on our planet.

3 Write each of these as ordinary numbers.

    a 3.65×10^4 **b** 1.725×10^6 **c** 6.075×10^5

    d 8.104×10^9 **e** 3.565×10^3 **f** 2.85×10^{12}

4 Write these numbers in standard index form.

    a 0.27   **b** 0.000031

    c 0.00000367   **d** 0.000001056

    e 0.0000000102   **f** 0.00207

    g 0.00000000105565   **h** 0.000020047

5 Write each of these as an ordinary number.

    a 2.62×10^{-4} **b** 1.05×10^{-7} **c** 3×10^{-8}

    d 7.65×10^{-11} **e** 8.45×10^{-9} **f** 5.375×10^{-7}

6 Use a calculator to find the following, giving your answers in standard form where appropriate.

    a A man walks 8.76×10^4 km in an 80-year lifetime.

        How many km does he walk on average each day?

    b The population of the UK is 6.1×10^7.

        The UK has a land area of 2.45×10^5 km^2.

        On average how many people live in each square kilometre in the UK?

    c The speed of light is about 3×10^5 km/s.

        The distance from the Sun to the Earth is about 1.5×10^8 km.

        About how long does light take to reach the Earth from the Sun?

7 Work out the value of each of the following.

a $\dfrac{6^4 \times 6^3}{6^5}$ b $\dfrac{3^2 \times 3^4}{3^3}$ c $10^4 \div 10^6$

d $7^3 \div 7^{-4}$ e $9^{-2} \times 9^{-4}$ f $\dfrac{4^3 \times 4^{-5}}{4^2}$

8 Simplify the following, leaving your answers as a single power of the number.

a $\dfrac{5^4 \times 5^3}{5^2 \times 5^3}$ b $\dfrac{6^2 \times 6 \times 6^7}{6^3 \times 6^4}$ c $\dfrac{8^3 \div 8^5}{8^2}$

d $\dfrac{8 \times 8 \times 8^2}{8^5}$ e $\dfrac{2^3 \div 2^4}{2^2}$ f $\dfrac{6^7 \times 6^{-2}}{6^3}$

9 Calculate the following, leaving your answers in index form.

a $(2^{-2})^2$ b $(3^2)^3$ c $(4^{-1})^{-2}$
d $(5^2)^{-3}$ e $(10^{-2})^{-5}$ f $(3^{-7})^{-2}$
g $((7^2)^{-3})^4$ h $((8^3)^3)^3$ i $((9^{-2})^{-2})^2$

10 Calculate the following, giving your answer to 2 dp where appropriate.

a $\dfrac{\sqrt{13^2 - 12^2} + 4}{5^2 - 4^2}$ b $\dfrac{(3^3 - 2^2)(9 + 4)^2}{5\sqrt{3^2 + 4^2}}$ c $\dfrac{\sqrt{2 + 0.4^2} + 5}{5 - \sqrt{2 + 0.4^2}}$

11 Write these numbers in their simplest form.

a $\sqrt{75}$ b $\sqrt{180}$ c $\sqrt{150}$
d $\sqrt{45}$ e $\sqrt{245}$ f $\sqrt{450}$

12 Calculate each of the following, leaving your answer in surd form.

a $3\sqrt{2} \times 5\sqrt{2}$ b $4\sqrt{3} \times 2\sqrt{3}$ c $2\sqrt{5} \times 4\sqrt{2}$
d $4\sqrt{7} \times 3\sqrt{3}$ e $2\sqrt{27} \times 3\sqrt{6}$ f $3\sqrt{12} \times 2\sqrt{8}$

13 Work out the value of each of the following.

a $4^{\frac{1}{2}}$ b $9^{-\frac{1}{2}}$ c $16^{\frac{3}{2}}$
d $64^{\frac{2}{3}}$ e $27^{-\frac{2}{3}}$ f $64^{-\frac{3}{4}}$

14 Calculate the following, leaving your answer in index form.

a $2^{\frac{2}{3}} \times 2^{\frac{1}{3}}$ b $3^{\frac{3}{4}} \times 3^{\frac{1}{2}}$ c $6^{\frac{1}{2}} \times 6^{\frac{1}{4}}$
d $4^{\frac{2}{3}} \div 4^{\frac{1}{2}}$ e $4^{-\frac{2}{3}} \div 4^{\frac{1}{3}}$ f $5^{\frac{3}{2}} \div 5^{-\frac{1}{3}}$

12 Constructions and Pythagoras

Introduction

How do you build a pyramid? The Great Pyramid of Giza is one of the seven wonders of the ancient world. It was completed in approximately 2560 BC and took about 20 years to build. At its time and for the next 3000 years it was the tallest building in the world!

What's the point?

To construct any complex building requires accurate scale drawings and mathematical calculations for all the dimensions.

Objectives

By the end of this chapter, you will have learned how to ...

- Understand and use Pythagoras' theorem.
- Use Pythagoras' theorem in real-life contexts.
- Construct a triangle with ruler and compasses.
- Draw the locus of a point from a given rule.

Check in

1 Find the missing numbers in these equations.

a $15^2 + 8^2 = \square^2$ b $5^2 + \square^2 = 13^2$

c $\sqrt{4.1^2 + 3.9^2} = \square$ (2 dp) d $7.2^2 + \square^2 = 9^2$

2 a Draw two lines, AB and AC, each 6 cm long with angle CAB = 60°.

b Using a ruler and compass only, construct the perpendicular bisector of AB and the angular bisector of CAB. Label the point where the two lines cross P.

c Measure **i** AP **ii** BP **iii** CP.

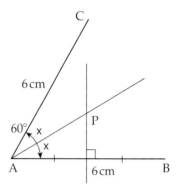

Starter problem

The Great Pyramid of Giza was built to a height of 146.5 m and has a square base of side 230.4 m.

Design and build an accurate scale model of the pyramid.

The longest side of a **right-angled triangle** is called
the **hypotenuse**.

You can draw squares on each side of a right-angled triangle.
The yellow area + the blue area = the green area

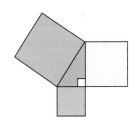

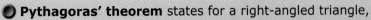

Pythagoras' theorem states for a right-angled triangle,
$$c^2 = a^2 + b^2 \quad \text{where } c \text{ is the hypotenuse.}$$
▶ the area of the square $=$ sum of the areas of the
on the hypotenuse squares on the other two sides.

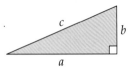

Example

a Calculate the value of c.

b Calculate the value of a.

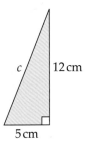

12 cm

5 cm

8 cm 10 cm

a

- - - - - - - - - -

a c is larger than 12 cm and 5 cm.

b a is smaller than 10 cm.

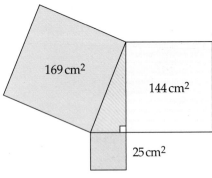

169 cm²

144 cm²

25 cm²

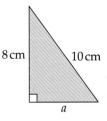

100 cm²

64 cm²

36 cm²

$c^2 = 12^2 + 5^2$ You add the areas.
$\quad = 144 + 25$
$\quad = 169$
$c = \sqrt{169}$ $\sqrt{}$ means **square root**.
$\quad = 13\text{ cm}$

$a^2 = 10^2 - 8^2$ You subtract the areas.
$\quad = 100 - 64$
$\quad = 36$
$a = \sqrt{36}$
$\quad = 6\text{ cm}$

Exercise 12a

1 Calculate the length of the hypotenuse in each triangle.
State the units of your answer.

a

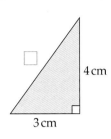

4 cm

3 cm

b

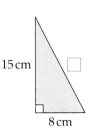

15 cm

8 cm

c

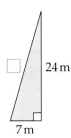

24 m

7 m

2 Calculate the unknown lengths in these right-angled triangles.

a

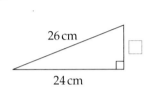

26 cm

24 cm

b

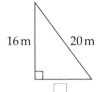

16 m   20 m

c

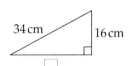

34 cm   16 cm

3 Calculate the unknown lengths in these triangles.

> Each answer is an integer.

a

12 m

9 m

b

8 cm   10 cm

c

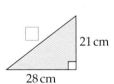

21 cm

28 cm

d

40 cm

9 cm

e

2.5 m

1.5 m

f

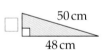

50 cm

48 cm

Problem solving

4 Farmer Giles has a field 250 m by 120 m. How much shorter is it from A to C across the diagonal than around the edge?

B   C

120 m

A   D

250 m

5 Draw a 3 cm, 4 cm, 5 cm right-angled triangle with a square on each side.

Cut out another 3 cm by 3 cm square.

Cut out another 4 cm by 4 cm square and cut it into 4 pieces exactly as shown.

Use these five pieces to fit in the 5 cm by 5 cm square to demonstrate Pythagoras' theorem.

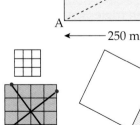

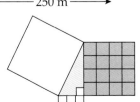

You can use Pythagoras' theorem to calculate the length of a side in a right-angled triangle.

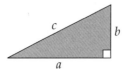

It must be a right-angled triangle!

● $c^2 = a^2 + b^2$ where c is the hypotenuse.

Example

Calculate the distance between the points (-3, 6) and (5, 2).

First draw the right-angled triangle.
The hypotenuse is c.

$c^2 = 4^2 + 8^2$

$= 16 + 64$

$= 80$

$c = \sqrt{80}$

$= 8.9$ units (1 dp)

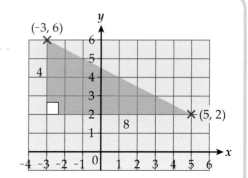

Example

Calculate the area of the equilateral triangle.

Use Pythagoras' theorem in the right-angled triangle.

$h^2 = 10^2 - 5^2$

$= 100 - 25$

$= 75$

$h = \sqrt{75}$

$= 8.660254...$ cm   Do not round the value of h.

p.246 >

Area of the equilateral triangle

$= \frac{1}{2} \times 10 \times 8.660254...$

$= 43.30127...$

$= 43.3$ cm^2 (1 dp)   Now round the answer.

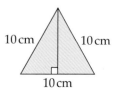

A **Pythagorean triple** consists of three integers a, b, c where $a^2 + b^2 = c^2$
Can you find any other Pythagorean triples?

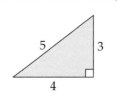

The most well-known Pythagorean triple is 3, 4, 5.

Exercise 12b

1 Use Pythagoras' theorem to decide if these triangles are right-angled.

a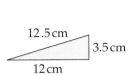
6 cm 7.5 cm
4.5 cm

b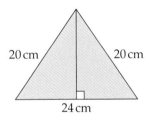
12.5 cm
3.5 cm
12 cm

c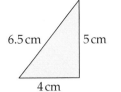
6.5 cm 5 cm
4 cm

d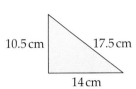
10.5 cm 17.5 cm
14 cm

2 Calculate the distances between these pairs of points.

   a (1, 2) and (4, 6) **b** (2, -1) and (3, 2) **c** (-2, 5) and (2, 1) **d** (-1, -2) and (1, 0)

Problem solving

3 A 4 metre ladder leans against a wall with its base
1.5 metres from the wall.
How far up the wall does the ladder reach?

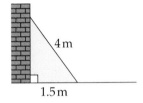
4 m
1.5 m

4 Draw a rectangle measuring 4 cm by 6 cm.

   a Draw and measure the diagonal of the rectangle.

   b Use Pythagoras' theorem to check your answer.

5 Two isosceles triangles have sides of length 24 cm, 20 cm, 20 cm
and 32 cm, 20 cm, 20 cm respectively.
For each triangle calculate its

   i perpendicular height   **ii** area.

a
20 cm 20 cm
24 cm

b
20 cm 20 cm
32 cm

Did you know?

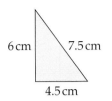

Euclid was a Greek mathematician living in Alexandria, Egypt around 300 BC. He is called the father of geometry.

6 The famous mathematician Euclid generated Pythagorean triples
(*a, b, c*) using the formulae

   $a = 2pq$ $b = p^2 - q^2$ $c = p^2 + q^2$

where *p* and *q* are integers and $p > q$.

   a Check that *a, b* and *c* form a Pythagorean triple when

      i $p = 2$ and $q = 1$ **ii** $p = 3$ and $q = 2$ **iii** $p = 4$ and $q = 3$

   b Investigate Pythagorean triples if *p* and *q*

      i are both even   **ii** are both odd   **iii** have a common factor greater than 1.

You can always **construct** a **unique** triangle if you are given

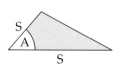

Two sides and the
included angle (SAS)

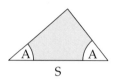

Two angles and the
included side (ASA)

A triangle is unique if it is the only one possible using the information given.

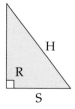

Three sides (SSS)

A right angle, hypotenuse and side (RHS)

The hypotenuse is the longest side in a right-angled triangle.

You cannot construct a unique triangle if you are given

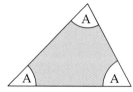

Three angles (AAA)

AAA: Many similar triangles can be constructed from the same information.

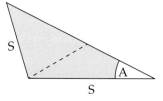

Two sides and the non-included angle (SSA)

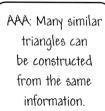

SSA: Two different triangles may be constructed from the same information.

Example

a Construct a triangle PQR where PQ = 7.5 cm, PR = 4 cm and angle Q = 25°.

b Can you draw more than one triangle with this information?

a First draw a sketch   then a construction.

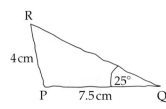

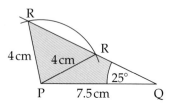

b Two possible triangles can be constructed with this information. There are two possible positions for the vertex R.

The information you are given in this example is SSA.

Exercise 12c

1 State whether the information given in each triangle is SAS, ASA, SSS or RHS.
Construct each triangle using ruler, protractor and compasses.

a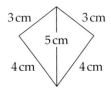

3.5 cm   4 cm
5 cm

b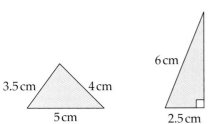

6 cm
2.5 cm

c

4.5 cm
140°
4 cm

d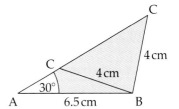

100°   50°
5 cm

2 a Construct the two different triangles that are possible for this information.
AB = 6.5 cm, BC = 4 cm, angle A = 30°

b Measure AC for each of your triangles.

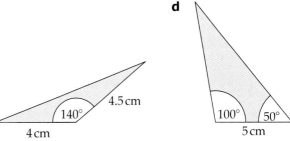

C

C   4 cm

4 cm

30°

A   6.5 cm   B

3 Construct these quadrilaterals.
Measure the length of the diagonals in each quadrilateral.

a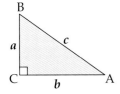

3 cm   3 cm
5 cm
4 cm   4 cm

b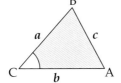

4 cm
7 cm   7 cm
4 cm

c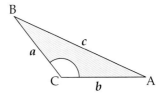

4 cm
100°   60°
4 cm   4 cm

Problem solving

4 You are asked to construct a triangle ABC, where angle A = 70° and angle B = 60°.
What one extra piece of information do you need to ensure you construct a unique triangle?

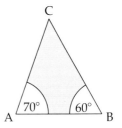

C

70°   60°
A   B

5 Repeat the construction of the RHS triangle in question **1** using only a ruler and compasses.

6 Angle C is either a right angle, an acute or an obtuse angle in these triangles.

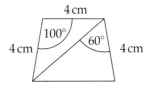

B
a   c
C   b   A

B
a   c
C   b   A

B
a   c
C   b   A

If angle C = 90°, then Pythagoras' theorem states $c^2 = a^2 + b^2$.
Construct triangles to investigate the relationship between c^2 and $a^2 + b^2$ if angle C is acute or if it is obtuse.

The athlete's hammer goes in a circle around his body.

The **locus** of an object is its path.

The plural of locus is loci.

> ● A point that moves according to a rule forms a locus.

The rule could be a set of instructions or a computer **program**.

▶ REPEAT 2
▶ [FORWARD 5 TURN RIGHT 60°
▶ FORWARD 5 TURN RIGHT 120°]

The locus is a rhombus.

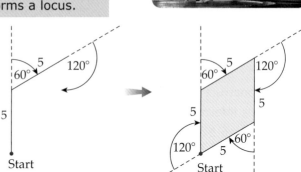

Example

Find the three loci of points that are the same distance from A and B, from B and C, and from C and A.

The set of points equidistant from A and B lie on the **perpendicular bisector** of the line segment AB.
Similarly for BC and CA.

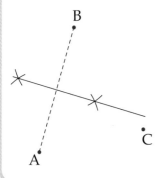 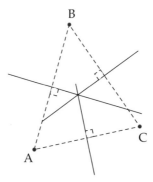

> The common point which is equidistant from A, B and C is the centre of the circumcircle of triangle ABC.

Example

A conger eel lives in a hole in an underwater cliff face. If a diver's finger comes within 50 cm of the hole it will be bitten. Describe the locus of the finger's closest safe approach.

The locus of points of closest safe approach form a **hemisphere** of radius 50 cm centred on the hole.

Exercise 12d

1 For the following sets of instructions
   i construct the shape using a suitable scale
   ii give the mathematical name of the shape produced.
   a REPEAT 3
      [FORWARD 6 TURN LEFT 120°]
   b FORWARD 8, RIGHT 120°
      REPEAT 3
      [FORWARD 4 TURN LEFT 60°]

2 Write a set of instructions for drawing
   a a regular pentagon   **b** a letter 'W'.

3 Three friends stand at the corners of a triangular field.
   Each friend walks towards the centre of the field being careful
   to stay an equal distance from the two edges of the field nearest
   them. By drawing your own triangle (about 8 cm across) and
   constructing the friends' paths, show that they will all pass
   through the same point.

The common point is the
centre of the inscribed circle.

4 A goat is tethered 3 m from the end of an 8 m long wall.
   On a scale drawing show the area that the goat can reach
   if the rope is
   a 2 m long   **b** 5 m long   **c** 6 m long.

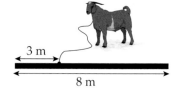

5 A Van de Graaff generator puts a large electrical charge on a
   metal sphere of diameter 20 cm. If your finger comes closer than
   30 cm you receive an electric shock. Describe the locus of points
   that are just at a safe distance.

6 A man stands 2 m from the top of a 6 m ladder which suddenly
   starts to slide down the wall. By making a scale drawing find
   the locus of the man's feet.

Problem solving

7 Mark two points A and B approximately 6 cm apart.
   Place a sheet of paper so that the edges pass
   through points A and B.
   Mark the position X, where angle AXB = 90°.
   Move the sheet of paper to different positions,
   each time marking the position of X,
   where angle AXB = 90°.

   Describe the locus of the point X.

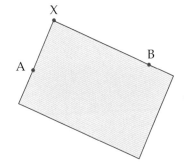

Check out

Test it ➡

You should now be able to ...

Questions

✓ Understand and use Pythagoras' theorem.	7	1 – 3
✓ Use Pythagoras' theorem in real-life contexts.	7	4, 5
✓ Construct a triangle with ruler and compasses.	7	6 – 8
✓ Draw the locus of a point from a given rule.	7	9

Language	Meaning	Example
Hypotenuse	The longest side of a right-angled triangle (Always opposite the right-angle).	hypotenuse
Pythagoras' theorem	$a^2 + b^2 = c^2$	13, 12, 5 $12^2 + 5^2 = 13^2$
Pythagorean triple	Any three integers a, b, and c that obey Pythagoras' theorem.	3, 4, 5 as $3^2 + 4^2 = 25 = 5^2$
Locus	The set of points that obey a given rule.	The 2D locus of all points 5 cm from A are in a circle radius 5 cm, centre A

1 Calculate the unknown lengths in these right angled triangles.

a

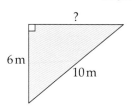

10 cm, 24 cm, ?

b

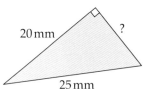

6 m, 10 m, ?

c

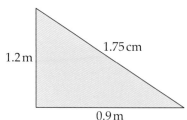

20 mm, 25 mm, ?

2 Use Pythagoras' theorem to decide whether this triangle is right-angled.

1.2 m, 1.75 cm, 0.9 m

3 Calculate the distances between these pairs of points. Give your answers as surds in their simplest form.

a (4, 8) and (-6, 13)

b (-5, -7) and (-9, 9)

4 A ladder of length 7 m leans against a wall with its base 2 m from the wall. How far up the wall does the ladder reach? Give your answer to the nearest cm.

5 What is the length of the diagonal of a 5 cm square? Give your answer in simplified surd form.

6 Construct this quadrilateral accurately.

a

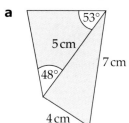

53°, 5 cm, 7 cm, 48°, 4 cm

7 Construct the *two* different triangles that are possible from this information.
AB = 5 cm, BC = 3 cm,
angle A = 25°

8 Write a set of instructions for drawing an equilateral triangle with side length 4.

9 A goat is tethered 2 m from the corner of a shed as shown in the picture. The rope is 3 m long.
On a scale drawing show the area that the goat can reach.

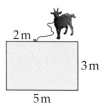

2 m, 3 m, 5 m

What next?

Score		
0 – 4		Your knowledge of this topic is still developing. To improve look at Formative test: 3C-12; MyMaths: 1112, 1090 and 1147
5 – 7		You are gaining a secure knowledge of this topic. To improve look at InvisiPen: 371, 375 and 381
8 – 9		You have mastered this topic. Well done, you are ready to progress!

12a

1 Use Pythagoras' theorem to calculate the unknown lengths.

a

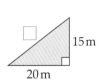

15 m

20 m

b

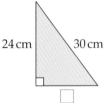

24 cm   30 cm

c

8.5 cm

7.5 cm

12b

2 Calculate the distance between these pairs of points.
   a (3, 4) and (6, 8)   **b** (3, -1) and (-1, 3)
   c (-7, 4) and (-2, -3)   **d** (1, 2) and (-8, -10)

3 A square is drawn inside a 4 by 4 square.
   Calculate
   a the length *c*
   b the area of the shaded square.

4 There are two right-angled triangles in the diagram.
   Calculate the value of *p* and of *q*.

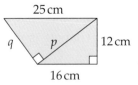

25 cm

q   *p*   12 cm

16 cm

12c

5 Using the information given
   i draw a sketch of each triangle
   ii state whether the information you are given is
       SAS, ASA, SSS, RHS, AAA or SSA.
   iii Does the information given result in a unique triangle?

a
| angle A = 80° |
| angle B = 70° |
| angle C = 30° |

b
| QR = 10 cm |
| PQ = 8 cm |
| PR = 9 cm |

c
| angle D = 30° |
| angle E = 40° |
| DE = 5 cm |

d
| JK = 8 cm |
| LK = 7.5 cm |
| angle K = 40° |

e
| angle S = 90° |
| ST = 3 cm |
| UT = 8 cm |

f
| YZ = 6 cm |
| XZ = 4 cm |
| angle Y = 35° |

6 Construct these shapes accurately.

a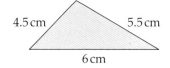

b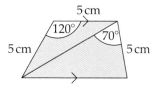

c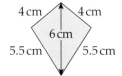

d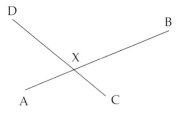

7 For the following sets of instructions
   i   construct the shape using a suitable scale
   ii give the mathematical name of the shape produced.

    a REPEAT 2
       [FORWARD 8 TURN RIGHT 90°
       FORWARD 5 TURN RIGHT 90°]

    b REPEAT 2
       [FORWARD 5 TURN RIGHT 45°
       FORWARD 3 TURN RIGHT 135°]

8 Two lines AB and CD intersect at X.

a Construct the locus of a point equidistant
    i from XA and XD   **ii** from XD and XB.
b What is the angle between these loci?

9 A radio mast can transmit a signal within a 40 km radius.
Draw a scale drawing to show the region that can receive the signal.
Use a scale of 1 : 10 000

10 PQR is a triangle with PQ = 7 cm,
PR = 9 cm and QR = 8 cm.
Shade in the region which is both less than
6 cm from R *and* closer to QR than QP.

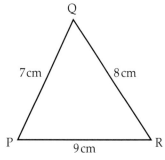

Sensory gardens are designed to stimulate the senses - sight, sound, smell, touch and even taste - and are thought to have a beneficial effect on people who visit them. While they must be designed for all users, this case study considers their accessibility for wheelchair users.

Raised flowerbeds are easier to reach for a person in a wheelchair.

Cross section of a raised bed

PLAN FOR A SENSORY CORNER

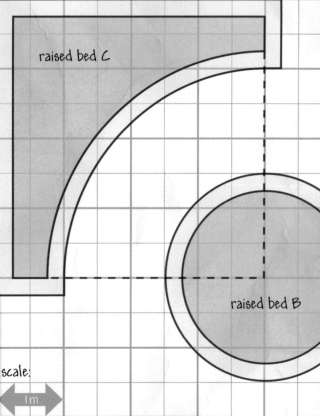

raised bed C

raised bed B

scale:

1m

Task 1

Look at the scale drawing of the garden.

Calculate the area in m² of

a) bed A   b) bed B (to 1 dp)

By considering different shapes, calculate the area of

c) bed C (to 1 dp)   d) bed D (to 1 dp)

Task 2

Look at the cross-section diagram of a raised bed. Each bed is to be filled with soil to 5cm from the top of the wall. Calculate the volume of soil needed to fill

a) bed A   b) bed B   c) bed C
d) bed D

Give your answers in m³ to 1 dp where appropriate.

Wide paths and few sharp corners make it easier to get around.

ts have different feels, different scents and e different sounds as the wind blows. They attract insects which add to the sounds.

Task 4

The area surrounding the beds and the path will be paved. Calculate the area that is to be paved, giving your answer in m² to 1 dp

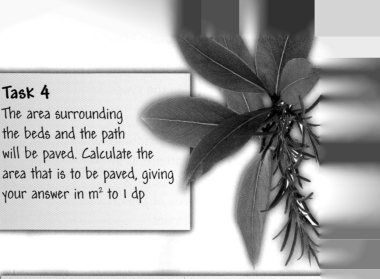

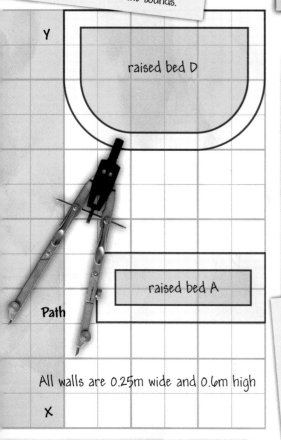

Y

raised bed D

raised bed A

Path

All walls are 0.25m wide and 0.6m high

X

Task 5

The path is made extra wide to fit a wheelchair comfortably.

a) Looking at the scale drawing, how wide is the path?

b) The path is to be sloped to provide access for wheelchair users.

It will have a gradient of 1:20, starting at X and rising up to Y.

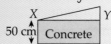

X ———— Y

50 cm | Concrete

i) At what height above bed B will the path be, at the point where the path meets the bed?

ii) At what height above bed D will the path be, at the end of the path Y?

c) (challenge) Find the total volume of concre needed for the path, giving your answer in m³ to 1 dp

Water features add sound and touch to a garden.

Task 3 (challenge)

Look again at the cross-section diagram of a raised bed.

Calculate the volume of concrete needed to make the **foundations** of

a) bed A   b) bed B

Calculate the volume of concrete needed to make the **walls** of

c) bed C   d) bed D

Give your answers in m³ to 1 dp

These questions will test you on your knowledge of the topics in chapters 9–12.
They give you practice in the types of questions that you may see in your GCSE exams.
There are 60 marks in total.

1 The diagram shows a quadrilateral ABCD.

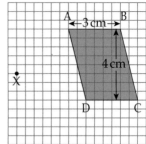

   a Enlarge the quadrilateral by scale
factor $\frac{1}{3}$ about the point X.   (3 marks)

   b Calculate the area of the original shape
and the enlargement.   (2 marks)

   c How are the areas related to the
scale factor?   (1 mark)

2 The diagram shows a
quadrilateral ABCD.

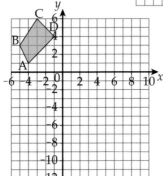

   a Enlarge the shape by
scale factor -2
using the origin as the
centre of enlargement.   (3 marks)

   b Write down the coordinates
of A'B'C'D'.   (3 marks)

3 A model kit of an aircraft is in the scale of 1 : 32.
The model aircraft has a wing span of 19 cm.
What is the wingspan on the real aircraft in metres?   (2 marks)

4 A 1 : 25 000 scale OS map shows the distance
between two points to be 15 cm.
Find the actual distance between these two points.   (2 marks)

5 The shapes A and B are similar.
B is 4 times larger than A.

   a Find the length marked a.   (1 mark)

   b Find the size of the angle marked b.   (1 mark)

   c In terms of area, how much bigger is B than A?   (1 mark)

6 Here is a pair of simultaneous equations.

$$2x - y = 4$$
$$x + y = 5$$

   a For each equation, tabulate the x and y values for x between 0 and 5.   (3 marks)

   b Draw the x-axis between -1 and 6 and the y-axis between -5 and 6.   (2 marks)

   c Plot the points for both straight lines and find the coordinates for the
point of intersection.   (4 marks)

7 Solve the inequalities

    a $3m - 6 \leq 9$ (2 marks)   **b** $5r - 4 < 11r - 16$ (2 marks)

8 The volume of this prism is $35\,\text{cm}^2$.

    a Formulate an expression for the
       volume of this prism.   (3 marks)

    b Use trial-and-improvement to find
       the value of x.   (4 marks)

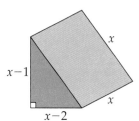

9 Write these numbers in standard form.

    a A wavelength of 450×10^{-9} km in metres.   (1 mark)

    b A radio frequency of $550\,\text{MHz}$ in Hertz.   (1 mark)

    c The distance to the Sun of $148\,800\,000\,\text{km}$ in metres.   (1 mark)

10 Use your calculator to evaluate this expression to 2 dp.   (2 marks)

$$\frac{\sqrt{(5.2^2 - 3.5)}}{7.3 + \sqrt{(0.5^2 + 0.4)}}$$

11 Calculate

    a $\sqrt{3} \times 6\sqrt{3}$ (1 mark)   **b** $5\sqrt{5} \times 3\sqrt{20}$ (1 mark)

12 Calculate the lengths marked with letters in each of these triangles.   (4 marks)
    Give your answers to two decimal places.

    a

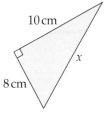

    b

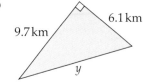

13 ABCD is a right-angled trapezium.

    a Work out the length of BC
       to two decimal places.   (3 marks)

    b Work out the length of AC
       to two decimal places.   (3 marks)

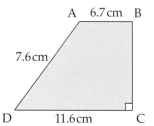

14 A right-angled triangle has side lengths $12\sqrt{8}\,\text{cm}$ and $10\sqrt{2}\,\text{cm}$.
    Work out the size of the hypotenuse.
    Leave your answer in the simplest surd form.   (4 marks)

13 Sequences

Introduction

There is a famous story told across many cultures of a wise man and a king. The wise man serves the king over many years and is rewarded with any prize he wishes to name. The man asks only for a chessboard and for one grain of rice to be placed on the first square, double this amount on the second square, double this amount on the third square and so on... The king agrees to this prize.
Was this a good prize to choose?

What's the point?

If the king had known a bit about sequences he would not have agreed to the prize!

Objectives

By the end of this chapter, you will have learned how to ...
- Find a position-to-term (nth) rule for a linear sequence.
- Explore triangular and square numbers.
- Find a position-to-term (nth) rule for a quadratic sequence.
- Explore the long-term behaviour of a sequence defined recursively.

Check in

1 For each of these sequences
  i write the term-to-term rule in words
  ii find the missing numbers.
  a 3, 6, 9, ☐, 15, ... **b** 2, 7, ☐, 17, 22, ... **c** 4, 7, ☐, ☐, 16, ...
  d 21, ☐, ☐, 9, 5, ... **e** 5, ☐, 21, ☐, 37, ... **f** 50, ☐, 38, ☐, 26, ...

2 Write the first five terms of each sequence, given its recursive formula.
  a $T(n + 1) = T(n) + 3$, $T(1) = 0$ **b** $T(n + 1) = 2T(n)$, $T(1) = 1$
  c $T(n + 1) = T(n) - 2$, $T(1) = 0$ **d** $T(n + 1) = \frac{1}{2}T(n)$, $T(1) = 2$

Starter problem

The cable for a suspension bridge is built using a hexagonal pattern for the parallel wire strands. Here are the first three layers.

$n = 1$ $n = 2$ $n = 3$

Investigate the number of wire strands in a cable for which $n = 12$.

13a Position-to-term rules

The **position-to-term** rule of a **sequence** allows you to find any term by substituting into a formula.

This sequence is **linear** because the difference between the terms is the same.

In the sequence 7, 14, 21, 28, 35, ...
the position-to-term rule or **nth term** is $T(n) = 7n$.

Example

Find the position-to-term rule of the sequence 2, 7, 12, 17, 22

2   7   12   17   22   The difference between consecutive
                           terms is 5 so the nth term involves
+5   +5   +5   +5   the 5 times table.

The nth term of a linear sequence is $T(n) = an + b$ where a is the difference between consecutive terms.

Position number, n	1	2	3	4	5
5 times table	5	10	15	20	25
nth term, $T(n)$	2	7	12	17	22

$\times 5$

-3

The nth term in words is 'multiply by 5 and subtract 3'
$T(n) = 5n - 3$
Check, $T(1) = 5 \times 1 - 3 = 2$ ✓,
$\qquad T(2) = 5 \times 2 - 3 = 7$ ✓

A pattern of diagrams may contain linear sequences.
▶ You can use diagrams to **justify** the form of the nth term.

Example

These pentagons are made using straws.

      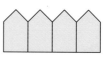

1 pentagon   2 pentagons   3 pentagons   4 pentagons

These are irregular pentagons.

The formula that relates the number of pentagons, p, to the number of straws, s, is

$\qquad s = 4p + 1.$

Justify this rule by referring to the diagrams.

The rule works because every new pentagon requires 4 straws plus 1 straw is needed to complete the first pentagon.

1 straw to complete the first pentagon.

4 straws for the first pentagon.

4 more straws for the second pentagon.

Exercise 13a

1 Generate the first five terms of a sequence whose nth term is given by

   a $T(n) = n + 10$ **b** $T(n) = 4n$ **c** $T(n) = n - 5$ **d** $T(n) = 2n + 3$

   e $T(n) = 5n - 1$ **f** $T(n) = \frac{1}{2}n + 1$ **g** $T(n) = 21 - n$ **h** $T(n) = 53 - 3n$

   i $T(n) = 30 - 2n$ **j** $T(n) = 6 - 4n$ **k** $T(n) = 3(n - 1)$ **l** $T(n) = 4(2n - 1)$

2 Find the nth term of each sequence.

   a 6, 12, 18, 24, 30, …   **b** 0, 1, 2, 3, 4, …   **c** 4, 7, 10, 13, 16, …

   d 7, 17, 27, 37, 47, …   **e** 36, 42, 48, 54, 60, …   **f** -8, -3, 2, 7, 12, …

   g 1.2, 1.4, 1.6, 1.8, 2.0, …   **h** $1\frac{1}{3}, 1\frac{2}{3}, 2, 2\frac{1}{3}, 2\frac{2}{3}$, …   **i** 49, 48, 47, 46, 45, …

   j 68, 61, 54, 47, 40, …   **k** -3, -8, -13, -18, -23, …   **l** $1\frac{1}{2}, 1, \frac{1}{2}, 0, -\frac{1}{2}$, …

Problem solving

3 Using the information given

   i find $T(n)$ **ii** generate the first five terms of the sequence.

   a $T(10) = 29$, $T(11) = 32$ and $T(12) = 35$

   b $T(8) = 42$ and $T(10) = 52$.

4 A flower bed is surrounded by square slabs.

The surround is always three slabs wide but it can be any length.

Make a table of values for different lengths of the flower bed from 3 to 7 slabs long and the total number of slabs used in each case.

Then find the general term for the sequence of the numbers of slabs.

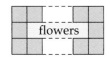

flowers

5 Jason makes a pattern of simple fish shapes using lollipop sticks.

 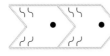 

   1 fish   2 fishes   3 fishes

   a Find a formula that connects the number of fishes, f, to the number of lollipop sticks, l.

   b Explain why this formula works.

   c Jason wants to make a wall frieze using 50 fishes.

       How many lollipop sticks will he need?

6 The position-to-term rule of a sequence is $T(n) = 2n + b$, where b is a constant.

   a Write the difference between consecutive terms of this sequence.

   b Write the value of b if

       i the first term is 0

       ii the sequence consists of the odd numbers, starting from 1.

   c Investigate other sequences of the form $T(n) = 3n + b$, $T(n) = 4n + b$, …

       $T(n) = an + b$.

MyMaths.co.uk   Q 1165   SEARCH

Here are the first four terms in the sequence of **square numbers**.

$T(1) = 1$ $T(2) = 4$ $T(3) = 9$ $T(4) = 16$

● The general term of the sequence of square numbers is $T(n) = n^2$

A triangular pattern of dots forms the sequence of **triangular numbers**.

$T(1) = 1$ $T(2) = 3$ $T(3) = 6$ $T(4) = 10$

● To find the nth triangular number add a row of n dots to the
$(n - 1)$th triangle.
▶ $T(n) = T(n - 1) + n$

1 dot
2 dots
3 dots
4 dots
5 dots $T(5) = 1 + 2 + 3 + 4 + 5 = 15$

To find the 5th triangular number add a row of 5 dots to the 4th triangle.

You can find the nth triangular number by evaluating the sum
$T(n) = 1 + 2 + 3 + \cdots + n$.
This formula becomes far too long when you want to find the 50th
or 100th triangular number!
You can find a general formula using rectangles.

$T(1) + T(1)$ $T(2) + T(2)$ $T(3) + T(3)$ $T(4) + T(4)$
$= 1 \times 2$ $= 2 \times 3$ $= 3 \times 4$ $= 4 \times 5$

The number of dots in the nth rectangle is $n(n + 1) = 2T(n)$

Each rectangle is made of two identical triangles. The number of dots in a rectangle is twice the triangular number.

● The general term of the sequence of triangular numbers is
$T(n) = \frac{1}{2}n(n + 1)$

Exercise 13b

1 For each sequence write the first five terms, the 10th term and 50th term.

   a $T(n) = 4n - 4$ **b** $T(n) = 4 - n$ **c** $T(n) = 100 - 4n$ **d** $T(n) = \dfrac{n^2}{2}$

   e $T(n) = n^2 - n$ **f** $T(n) = \dfrac{10}{n}$ **g** $T(n) = n - 2n^2$ **h** $T(n) = 100 - n^2$

2 The formula for the sum of the whole numbers from 1 to n is
   $T(n) = \dfrac{1}{2}n(n + 1)$.

   Use this formula to find the sum of the whole numbers from 1 to

   a 10   **b** 20   **c** 50   **d** 100   **e** 1000.

Problem solving

3 **a** Draw the next two diagrams in this sequence.

   $T(1) = 1$ $T(2) = 1 + 3$ $T(3) = 1 + 3 + 5$

   The nth term for this sequence in words is
   'sum the consecutive odd numbers from 1 to $2n - 1$'
   as a formula $T(n) = 1 + 3 + 5 + 7 + \cdots + (2n - 1)$.

   b By looking at the shape of the diagrams find another
        way of writing the nth term of this sequence.

   c Use your answer to part **b** to find the sum of the odd
        numbers from 1 to 49.

4 Write a formula that relate the height of the pile, h, to the
   number of bricks, b.

Did you know?

Since Pythagoras, people have been fascinated by numbers arising from patterns of objects. The square-based pyramid gives the sequence 1, 5, 14, 30, 51, 91, …

$T(n) = \dfrac{1}{6}n(n + 1)(2n + 1)$.

5 The *maximum* number of times that two straight lines can cross is
   once. The *maximum* number of times that three straight lines can
   cross is three.

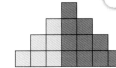

   2 lines   3 lines
   1 crossing   3 crossings

   a Draw diagrams for four and five straight lines.
   b Predict how the sequence continues.
   c Explain why the sequence continues in this way.

 MyMaths.co.uk

In the **quadratic** sequence

the first difference increases

but the **second difference** is constant.

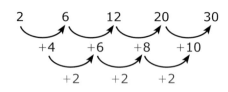

2   6   12   20   30

+4   +6   +8   +10

+2   +2   +2

The difference may increase or decrease.

> ● A sequence is linear if the first difference is constant, and quadratic if the second difference is constant.

A coefficient is a number that multiplies an algebraic term.

> ● The position-to-term rule of a quadratic sequence is always of the form
> $$T(n) = an^2 + bn + c$$
> where the **coefficients** a, b and c are constants and $b \neq 0$.

> ● The coefficient a is always half the value of the second difference.

You can use these facts to find the rule for a quadratic sequence.

Example

Find the position-to-term rule of the sequence 3, 11, 23, 39, 59, …

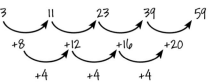

3   11   23   39   59

+8   +12   +16   +20

+4   +4   +4

The second difference is 4, so $a = \frac{1}{2} \times 4 = 2$. The formula will contain $2n^2$

Position number, n	1	2	3	4	5
Sequence, $T(n)$	3	11	23	39	59
Quadratic term, $2n^2$	2	8	18	32	50
Linear part, $T(n) - 2n^2$	1	3	5	7	9

+2   +2   +2   +2

Subtracting the quadratic term from the sequence leaves a linear sequence.

2 times table, $2n$	2	4	6	8	10
Subtract 1, -1	-1	-1	-1	-1	-1
Linear part, $2n - 1$	1	3	5	7	9

The difference is always 2 so the nth term of the linear part involves the 2 times table.

The **nth term** is $T(n) = 2n^2 + 2n - 1$.

Exercise 13c

1 Generate the first five terms of each quadratic sequence if

  a $T(n) = n^2$ **b** $T(n) = 3n^2$ **c** $T(n) = 2n^2 - 1$

  d $T(n) = n^2 - 5$ **e** $T(n) = 10 - 2n^2$ **f** $T(n) = n^2 + 2n$

  g $T(n) = 2n^2 - 3n$ **h** $T(n) = n^2 + 3n - 5$ **i** $T(n) = n^2 - 4n + 10$

  j $T(n) = 2n^2 + n - 4$ **k** $T(n) = 4n^2 - 2n + 1$ **l** $T(n) = \dfrac{n(n - 1)}{2}$

2 Complete the differences and nth terms for these quadratic sequences.

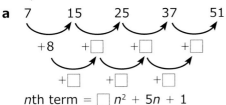

  a   7   15   25   37   51

       $+8$ $+\square$ $+\square$ $+\square$

       $+\square$ $+\square$ $+\square$

     nth term $= \square\, n^2 + 5n + 1$

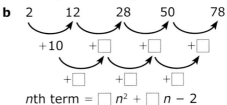

  b   2   12   28   50   78

       $+10$ $+\square$ $+\square$ $+\square$

       $+\square$ $+\square$ $+\square$

     nth term $= \square\, n^2 + \square\, n - 2$

3 Find the nth term of each of these quadratic sequences.

  a   2, 8, 18, 32, 50, …   **b**   5, 20, 45, 80, 125, …   **c**   2, 5, 10, 17, 26, …

  d   -1, 2, 7, 14, 23, …   **e**   4, 10, 18, 28, 40, …   **f**   7, 18, 33, 52, 75, …

  g   0, 1, 4, 9, 16, …   **h**   6, 15, 28, 45, 66, …   **i**   5, 13, 27, 47, 73, …

Problem solving

4 Aliens from the planet Zorg have a pattern of dots on their chest which relates to how old they are.

  a Copy and complete this table of values.

Age of alien in years, n	1	2	3
Number of dots, $T(n)$			

1 year old   2 years old   3 years old

  b Find the rule that connects the age of the alien in years, n, to the number of dots on their chest, $T(n)$.

  c Work out the number of dots on an alien aged 100 years.

  d King Zorgon states that he uses the formula $T(n) = n(n + 1)$. Explain why this formula works and prove that it is equivalent to the formula you found in part **a**.

5 A quadratic sequence has the general term $T(n) = an^2 + bn + c$.

  a Write an expression for $T(1)$ by substituting $n = 1$ into the formula.

  b Write expressions for $T(2)$, $T(3)$ and $T(4)$.

  c Using your knowledge of algebra, write the differences between consecutive terms in this sequence.

  d Find the second difference and hence prove that the second difference in a quadratic sequence is always $2a$.

The **long-term behaviour** of a sequence can be quite interesting!

Example

For each **recursive** formula, write out the first eight terms of the sequence and describe its behaviour.

a $T(n + 1) = 4T(n) + 1$ $\quad$ $T(1) = 0$

b $T(n + 1) = \frac{1}{2}T(n) + 1$ $\quad$ $T(1) = 0$

c $T(n+1) = -\frac{1}{2}T(n)$ $\quad$ $T(1) = 1$

---

a 0, 1, 5, 21, 85, 341, 1365, 5461
The numbers appear to be getting larger

b 0, 1, 1.5, 1.75, 1.875, 1.9375, 1.96875, 1.984375
The numbers are getting larger, but never seem to reach 2

c 1, −0.5, 0.25, −0.125, 0.0625, −0.03125, 0.015625, −0.0098125
The numbers are getting smaller, but keep switching from positive to negative

A **convergent sequence** approaches a **limiting value**.

A **divergent sequence** does not converge – it typically gets larger and larger.

An **oscillatory sequence** increases and decreases in turn; it could converge or diverge.

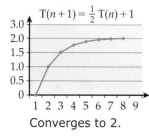

$T(n+1) = \frac{1}{2}T(n) + 1$

Converges to 2.

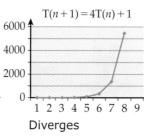

$T(n+1) = 4T(n) + 1$

Diverges

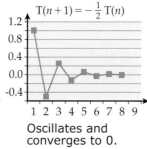

$T(n+1) = -\frac{1}{2}T(n)$

Oscillates and converges to 0.

Sequences can describe what happens in the real world.

Example

A radioactive substance decays annually according to the recursive formula $T(n + 1) = \frac{7}{8}T(n)$. There is initially 1 kg of the substance.

a How many years will it take for half of the substance to decay?

b How much substance will be left after 50 years?

---

a After 1 year, $1000 \times \frac{7}{8} = 875\,g$ $\quad$ Then 766, 670, 586, 513, 449, ...
The substance will have decayed to half its original amount sometime during the sixth year.

b $1000 \times \left(\frac{7}{8}\right)^{50} = 1.26\,g$ $\quad$ Use the index key on your calculator.

Did you know?

The time for half of a radioactive substance to decay is called its **half-life**. For plutonium, this is over 24 000 years.

Exercise 13d

1 Write down the first six terms of each sequence, and state if any are convergent.
In each case, take T(1) = 0.

   a $T(n + 1) = 3T(n) + 1$ **b** $T(n + 1) = \frac{1}{2}T(n) - 2$ **c** $T(n + 1) = \frac{4}{5}T(n) - 5$

   d $T(n + 1) = 1 - 2T(n)$ **e** $T(n + 1) = \{T(n)\}^2 - 2$ **f** $T(n + 1) = \{T(n)\}^2 + 3$

2 By working out the first few terms, describe the long-term behaviour of each sequence.
In each case, take T(1) = 0.

   a $T(n + 1) = 3T(n) - 1$ **b** $T(n + 1) = \frac{1}{2}T(n) + 1$ **c** $T(n + 1) = -2T(n) + 2$

   d $T(n + 1) = -\frac{1}{2}T(n) + 2$ **e** $T(n + 1) = \{T(n)\}^2 - \frac{1}{2}$ **f** $T(n + 1) = \{T(n)\}^2 + \frac{1}{2}$

3 Match the sketch graphs with the descriptions **A** to **D**.

   A   convergent   **B**   divergent

   C   convergent and oscillatory   **D**   divergent and oscillatory

   a   **b**   **c**   **d**

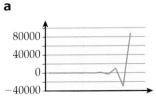 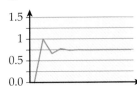 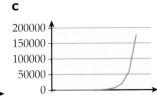 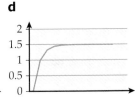

Problem solving

4 A forest fire is expanding at a rate given by the recursive formula

        $T(n + 1) = 2T(n) - 90$, with $T(1) = 100$

    where the units are kilometres squared per day.

   a   Find the area affected after 8 days.

   b   After 10 days, the fire starts to decrease at a rate of

        $T(n + 1) = \frac{1}{2}T(n) + 90$

    According to this formula, what area of forest will still be left burning long-term?

5 **a**   Choose any two whole numbers between 1 and 10, let your numbers be T(1) and T(2).
Now apply the recursive formula $T(n + 1) = T(n) + T(n - 1)$ to generate a sequence.

   b   Look at the ratio between successive terms, giving your answer
as a decimal to 3 dp.

   c   Write your ratios as a sequence. What can you say about its limit?

   d   Try again with any other two integers of your choice.

6 **a**   Investigate the limits of sequences of the form $T(n + 1) = kT(n)$, where k can be any
constant number of your choice (try positives, negatives, decimals less than 1 …).
Write a brief report of your findings.

   b   Repeat with sequences of the form $T(n + 1) = k\{1 - T(n)\}$

Check out

You should now be able to ...

Test it ➡

Questions

✓ Find a position-to-term (*n*th) rule for a linear sequence.	6	1, 2
✓ Explore triangular and square numbers.	7	3, 4
✓ Find a position-to-term (*n*th) rule for a quadratic sequence.	8	5, 6
✓ Explore the long-term behaviour of a sequence defined recursively.	8	7

Language	Meaning	Example
Position-to-term rule (*n*th term)	A rule that links each to its position in the sequence.	For the sequence 2, 4, 6, 8 the position-to-term rule 'multiply the position by 2' can be written $2n$.
Quadratic sequence	A sequence generated by a rule including a squared term.	1, 4, 9, 16, ... is generated by *n*th term n^2.
Triangular numbers	Numbers that can be shown as a triangular pattern of dots.	The first five triangular numbers are 1, 3, 6, 10, 15.
Recursive formula	Allows you to calculate any term of a sequence given the previous term.	$T(n + 1) = T(n) + 3$; $T(1) = 3$ generates the 3 times table.
Long-term behaviour	How a sequence looks after many terms.	Common behaviours are convergent, divergent and oscillatory.
Convergent	A sequence that gets closer and closer to a *limiting value*.	$1, -\frac{1}{2}, \frac{1}{4}, -\frac{1}{8}, \frac{1}{16}, -\frac{1}{32}$ converges to 0.
Divergent	A sequence that does *not* converge.	1, 2, 4, 8, 16, 32 1, -1, 1, -1, 1, -1
Oscillatory	A sequence in which alternate first differences change sign.	1.1, 0.99, 1.001, 0.9999, 1.00001 1, -2, 4, -8, 16, -32

1 Find the *n*th term for each of these sequences.
 a 1, 6, 11, 16, …
 b 4, 11, 18, 25, …
 c 5.6, 6.5, 7.4, 8.3, …
 d 18, 16, 14, 12, …

2 Generate the first 5 terms of these sequences.
 a $T(n) = 6n - 3$
 b $T(n) = \frac{1}{2}n - 5$
 c $T(n) = 4(6n + 1)$
 d $T(n) = 22 - 3n$.

3 The formula $T(n) = \frac{1}{2}n(n + 1)$ gives the sum of the first *n* whole numbers.
 a Use it to find the sum of the first 30 whole numbers.
 b Use it to find the 40th triangle number.

4

 a By considering the rows and columns find the *n*th term of this sequence.
 b How many squares will be in the 10th term?

5 Generate the first 5 terms of these quadratic sequences.
 a $T(n) = n^2 + 1$
 b $T(n) = 4n^2$
 c $T(n) = 3n(2n - 3)$
 d $T(n) = n^2 + 3n + 1$

6 Find the *n*th term of these quadratic sequences.
 a 6, 9, 14, 21, 30, …
 b 5, 14, 27, 44, 65, …
 c 3, 13, 31, 57, 91, …
 d -2, -4, -4, -2, 2

7 By working out the first few terms, describe the long-term behaviour of each sequence.
 In each case, take $T(1) = 0$.
 a $T(n + 1) = 2T(n) - 1$
 b $T(n + 1) = \frac{1}{2}T(n) - 1$
 c $T(n + 1) = \{T(n)\}^2 + 1$
 d $T(n + 1) = \{T(n)\}^2 - 1$

What next?

Score		
	0 – 3	Your knowledge of this topic is still developing. To improve look at Formative test: 3C-13; MyMaths: 1165 and 1166
	4 – 6	You are gaining a secure knowledge of this topic. To improve look at InvisiPen: 273 and 283
	7	You have mastered this topic. Well done, you are ready to progress!

13a

1 For sequences with the following nth terms, find
   i T(10)   **ii** T(100)
   a $T(n) = n + 5$ **b** $T(n) = 5n + 3$ **c** $T(n) = 6n - 3$
   d $T(n) = 2n - 9$ **e** $T(n) = 10 - 2n$ **f** $T(n) = 2(3n - 2)$

2 The position-to-term rule of a sequence is $T(n) = an + b$,
  where a and b are constants.
  a Write the value of a if the sequence is increasing in steps of 5.
  b What can you say about a if the sequence is decreasing in equal steps?

3 Find the nth term of each of these sequences.
  a 3, 8, 13, 18, 23, …   **b** -5, -2, 1, 4, 7, …   **c** 29, 28, 27, 26, 25, …
  d 12, 6, 0, -6, -12, …   **e** -8, -5, -2, 1, 4, …   **f** $2\frac{2}{3}, 2\frac{1}{3}, 2, 1\frac{2}{3}, 1\frac{1}{3}, …$

4 A kitchen wall has three rows of tiles.
  The two end columns have tiles with floral patterns.
  There are n central columns of plain white tiles.
  a Find the total number of tiles used when $n = 3$.
  b Find an expression $T(n)$ for the total number of tiles.
  c Write the sequence of the values of $T(n)$ for $n = 40$ to $n = 45$.

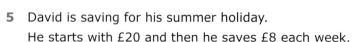

5 David is saving for his summer holiday.
  He starts with £20 and then he saves £8 each week.
  a How much has he saved after
    i 2 weeks   **ii** 5 weeks?
  b Find an expression for $T(n)$, the amount, in pounds, he
    has saved after n weeks.
  c How much has he saved after 20 weeks?

13b

6 Mr Stewart stacks tins in these formations.

  Find a formula that connects the height of each
  stack, h, to the number of tins, t, in the stack

7 The sum of two consecutive triangular numbers is always a square number.
  a Use diagrams to test this statement for particular cases.
  b Copy and complete this statement:
    '$T(n - 1) + T(n) = \square$ where $T(n)$ is the nth triangular number.'

8 Decide if each sequence is linear, quadratic or neither.

  a 2, 6, 12, 20, 30, …   **b** 20, 18, 16, 14, 12, …
  c 1, 8, 27, 64, 125, …   **d** 16, 9, 14, 21, 30, …
  e 0, 3, 8, 15, 24, …   **f** 1, 2, 4, 8, 16, …

9 Find the nth term of each of these quadratic sequences.

  a 3, 12, 27, 48, 75, …   **b** 3, 6, 11, 18, 27, …
  c 5, 12, 21, 32, 45, …   **d** 1, 6, 15, 28, 45, …
  e 3, 12, 25, 42, 63, …   **f** 1, 7, 17, 31, 49, …
  g 6, 16, 30, 48, 70, …   **h** 5, 19, 39, 65, 97, …
  i 2, 2, 4, 8, 14, …   **j** 9, 16, 25, 36, 49, …

10 Each dress has an embroidered
pattern of squares that relates
to its size.

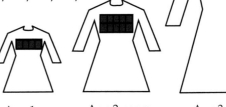

Age 1 year   Age 2 years   Age 3 years

  a Copy and complete this table of
  values.

Dress size (age in years), n	1	2	3
Number of squares, T(n)			

  b Find the rule that connects the dress size, n, to the number
  of embroidered squares, T(n).
  c The manufacturer of this dress state that they use the
  formula T(n) = $n(n + 3)$.
  Explain why this formula works and prove that it is
  equivalent to the formula you found in part **b**.

11 Workout the first six terms and describe the long-term behaviour of each sequence.

  a T(1) = 4,   T($n + 1$) = T(n) + 3
  b T(1) = 16,   T($n + 1$) = $\frac{1}{4}$T(n) − 2
  c T(1) = 2,   T($n + 1$) = 6 − T(n)
  d T(1) = 3,   T($n + 1$) = {T(n)}^{2} − 4

12 Repeat question **11**, taking T(1) = 1 for each sequence.
What do you notice about the long-term behaviour of each sequence?

13 A substance is eroding at a rate given by the recursive formula
T($n + 1$) = $\frac{2}{3}$ T(n) + 100, with T(1) = 1000 where the units are grams per month.

  a How many grams of the substance are left after 12 months?
  b What happens to the substance in the long term?

14 3D shapes and trigonometry

Introduction

Small animals such as meerkats only eat a few grams of food a day. However it would only take about a week for a meerkat to eat the equivalent of its own body mass in food. A human on the other hand, eats about 1kg of food a day but would take nearer two months to eat the equivalent of its own body mass in food. Meerkats have a greater surface area in proportion to their volume than humans, meaning they lose relatively more heat, meaning they need to eat relatively more.

What's the point?

The size and shape of different animals determines their metabolism. Understanding surface area and volume helps us to understand the animal kingdom.

Objectives

By the end of this chapter, you will have learned how to ...

- ● Classify 3D shapes and draw 2D representations.
- ● Calculate the surface area and volume of a prism.
- ● Use Pythagoras' theorem in three dimensions.
- ● Use sine, cosine and tangent to find lengths and angles in right-angled triangles.
- ● Use trigonometry in calculations with bearings.

Check in

1 State the mathematical names of these 3D shapes.

a  **b** 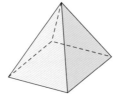 **c** 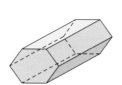 **d**

2 Calculate the length of the unknown sides in these right-angled triangles.

a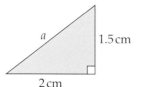

a 1.5 cm

2 cm

b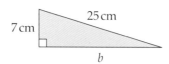

7 cm   25 cm

b

Starter problem

How do you measure the height of a mountain?

Have a look at this diagram.

Could you work out the height from this information?

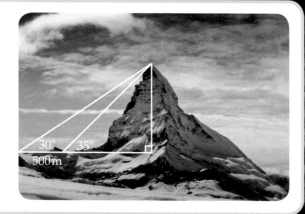

30° 35°
500 m

A solid is a shape formed in three-dimensions (3D).

You can describe a solid using projections of the 3D shape.
These projections are called

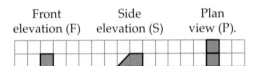

Front elevation (F) Side elevation (S) Plan view (P).

The plan is the 'birds eye view'. You can draw the solid using isometric paper.

A **net** is a 2D shape that folds to form a solid.

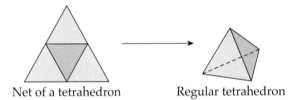

Net of a tetrahedron Regular tetrahedron

Some solids have reflection symmetry.

A line of symmetry divides a 2D shape into two halves that are mirror images.

A **plane of symmetry** divides a 3D shape into two halves, each of which is the mirror image of the other.

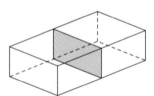

Example

This is the net of a solid.

a State the mathematical name of the solid.

b Draw diagrams to show any planes of symmetry.

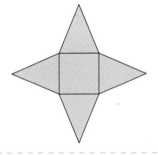

---

a A square-based pyramid.

b

A square-based pyramid has four planes of symmetry.

Exercise 14a

1 For each shape sketch the plan, front and side elevation

a 　　b 　　c 　　d

e 　　f 　　g 　　h

2 A regular octahedron is made from eight equilateral triangles.
   The solid has five planes of symmetry.
   a Describe the 2D shape formed when you slice the
      octahedron through each plane of symmetry.
   b Draw the front elevation (F), the side
      elevation (S) and the plan view (P).
   c Draw a net of the octahedron.

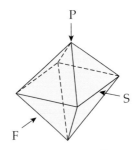

3 a State the number of faces, edges and vertices of this prism.
   b Draw a net of the prism.

4 On isometric paper, draw the solid that has these
   2D projections.

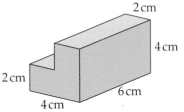

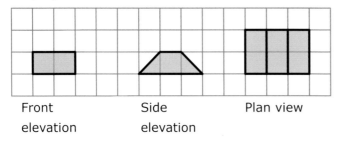

Front   Side   Plan view
elevation   elevation

Problem solving

5 The centre point of each face of a solid is marked with a dot.
   Straight lines are drawn from each dot to other dots on adjacent faces.
   State the mathematical name of the 3D shape generated by
   these lines if the solid is
   a a cube　　　b a regular tetrahedron.

6 a Construct a cube with edges of length 5 cm.
   b Use your model to decide the number of
      planes of symmetry of a cube.

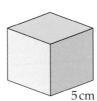

You calculate the **surface area** of a **prism** by finding the area of
each surface.

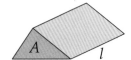

> ● **Volume** of a prism = area of **cross-section** × length
> = $A \times \ell$.

Example

A cylinder has a radius of 5 cm and a height of 6 cm. (Use π = 3.14)
Calculate   **a** the surface area
                   b the volume of the cylinder.

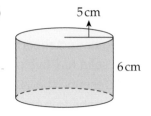

a

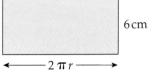

6 cm

The curved surface unrolls
to form a rectangle.

$\longleftarrow 2\pi r \longrightarrow$

The circumference of
a circle is $2\pi r$.

Area of the rectangle = $2\pi r \times 6$
                      = $2 \times \pi \times 5 \times 6 = 60\pi$
Area of one circular end = $\pi r^2 = \pi \times 5^2 = 25\pi$
          Surface area = $60\pi + 25\pi + 25\pi$
                      = 110π The answer in terms of π is exact.
                      = $345.4\,cm^2$ (1 dp)
b   Volume = area of circular cross-section × length
             = $25\pi \times 6$
             = 150π
             = $471\,cm^3$

Example

A cuboid measures
3 cm by 4 cm by 12 cm.
Calculate the length of
the diagonal AD.

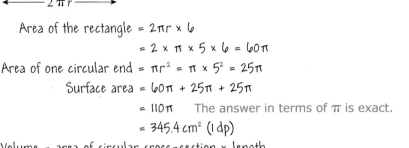

You can only
use Pythagoras'
theorem in right-
angled triangles.

❮ p.214

In triangle BCD
$BD^2 = 3^2 + 4^2$
     $= 9 + 16$
$BD^2 = 25$
$(BD = \sqrt{25} = 5\,cm)$
$AD^2 = AB^2 + BC^2 + CD^2$

In triangle ABD
$AD^2 = 5^2 + 12^2$
     $= 25 + 144$
     $= 169$
$AD = \sqrt{169} = 13\,cm$

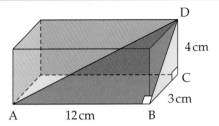

Exercise 14b

1 A cube has sides of length 16 cm.
   Calculate the surface area and the volume of the
   shaded prism.

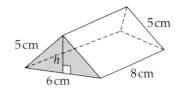

2 A prism has a right-angled
   triangle as the cross-section.
   Calculate
   a the length AC
   b the area of triangle ABC
   c the total surface area of the prism
   d the volume of the prism.

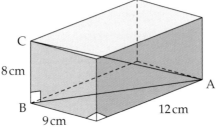

3 The diagram shows a triangular prism.
   Calculate
   a the height *h*
   b the surface area
   c the volume.

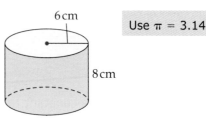

4 A cylinder has a radius of 6 cm and a height of 8 cm.
   Calculate
   a the surface area
   b the volume of the cylinder.

   Use π = 3.14

5 A cuboid measures 8 cm by 9 cm by 12 cm.
   Calculate
   a the length AB
   b the length CA.

Problem solving

6 Calculate the length of the longest straight rod that will fit
   completely inside a cubical box of side length 10 cm.

7 A cylinder has radius *r* and height *h*.
   a In terms of *r*, *h* and π, calculate
      i the surface area and
      ii the volume of the cylinder.
   b Check your formulae have the correct dimensions.

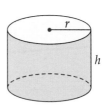

‹ p.20

You can name the sides in a right-angled triangle.

> ● The **hypotenuse** is the longest side of the triangle.
> ● The **opposite** side faces the angle marked θ.
> ● The **adjacent** side is beside the angle marked θ.

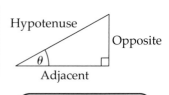

θ is the Greek letter theta.
The green triangle and the pink triangle are **similar**. The scale factor of the **enlargement** is s.

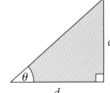

Two triangles are similar if two pairs of angles are equal.

$a \times s = c$ and $b \times s = d$

Dividing these equations gives

$$\frac{a}{b} = \frac{c}{d} = \frac{\text{opposite}}{\text{adjacent}}$$

For all similar right-angled triangles, the ratio $\dfrac{\text{opposite}}{\text{adjacent}}$ is constant.

This ratio is called the **tangent**.

> ● $\tan \theta = \dfrac{\text{opposite side}}{\text{adjacent side}}$

tan is the abbreviation for tangent.
sin is short for sine.
cos is short for cosine.

You can find two more constant ratios in a right-angled triangle.

> ● $\sin \theta = \dfrac{\text{opposite side}}{\text{hypotenuse}}$ $\cos \theta = \dfrac{\text{adjacent side}}{\text{hypotenuse}}$

Calculate the value of $\sin \theta$, $\cos \theta$ and $\tan \theta$ in this right-angled triangle.

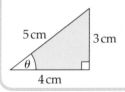

$\sin \theta = \dfrac{\text{opposite}}{\text{hypotenuse}} = \dfrac{3}{5} = 0.6$

$\cos \theta = \dfrac{\text{adjacent}}{\text{hypotenuse}} = \dfrac{4}{5} = 0.8$

$\tan \theta = \dfrac{\text{opposite}}{\text{adjacent}} = \dfrac{3}{4} = 0.75$

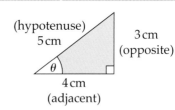

A mnemonic is

SOH   CAH   TOA

$S = \dfrac{O}{H}$ $C = \dfrac{A}{H}$ $T = \dfrac{O}{A}$

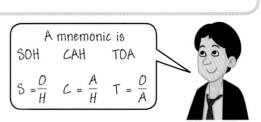

Exercise 14c

1 Find the missing lengths in these pairs of similar triangles.

a
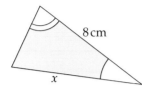

3 cm
4 cm
8 cm
x

b
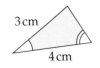
3 cm
y
6 cm
7 cm

c
8 cm
z
6 cm
7 cm

2 For each right-angled triangle, calculate as a decimal to 3 significant figures, the values of sin θ, cos θ and tan θ. All measurements are in centimetres.

a
5 θ 3
4

b
13
5
θ
12

c
15   17
8   θ

d
θ
24   25
7

Problem solving

3 **a** Explain why these triangles are similar.
b Accurately construct each triangle.
c Measure the lengths a, b, c and d.
d Calculate the values of $\dfrac{a}{b}$ and $\dfrac{c}{d}$.
e Which trigonometric ratio have you found?

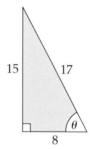
b a
$60°$
3 cm

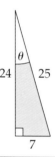
d c
$60°$
5 cm

4 ABC is an equilateral triangle.
Calculate the value of cos 60°.

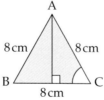
A
8 cm   8 cm
B   C
8 cm

5 On graph paper draw a quadrant of a circle with radius 10 cm.
Use a protractor to construct a triangle OAB so that angle AOB = 50°.
Measure the length AB.
Compare your answer for $\dfrac{AB}{10}$ with the calculator value of sin 50°. Repeat the process for other angles.
What do you notice?

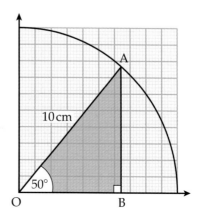
A
10 cm
50°
O   B

You use **trigonometric** ratios to calculate an angle or a side in a right-angled triangle.

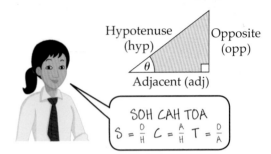

- $\sin \theta = \dfrac{\text{opposite}}{\text{hypotenuse}}$ $\cos \theta = \dfrac{\text{adjacent}}{\text{hypotenuse}}$

  $\tan \theta = \dfrac{\text{opposite}}{\text{adjacent}}$

SOH CAH TOA
$S = \frac{O}{H}$ $C = \frac{A}{H}$ $T = \frac{O}{A}$

Example

Calculate the angle θ in each right-angled triangle.

a

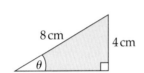
8 cm   4 cm

b

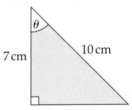
7 cm   10 cm

a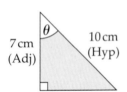
(Hyp)
8 cm   4 cm
(Opp)

Sin uses opposite and hypotenuse.

$\sin \theta = \dfrac{\text{opp}}{\text{hyp}} = \dfrac{4}{8} = 0.5$

$\theta = \sin^{-1} 0.5 = 30°$

Use the inverse buttons on your calculator.

b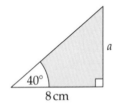
7 cm
(Adj)   10 cm
(Hyp)

Cos uses adjacent and hypotenuse.

$\cos \theta = \dfrac{\text{adj}}{\text{hyp}} = \dfrac{7}{10} = 0.7$

$\theta = \cos^{-1} 0.7 = 45.6° \ (1 \text{dp})$

Example

Calculate the lengths *a* and *b* in these right-angled triangles.

a

a
40°
8 cm

b

5 cm
30°
b

a
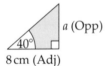
a (Opp)
40°
8 cm (Adj)

Tan uses opposite and adjacent.

$\tan 40° = \dfrac{\text{opp}}{\text{adj}} = \dfrac{a}{8}$

$a = 8 \times \tan 40° = 6.7 \text{cm} \ (1 \text{dp})$

Rearrange the equation to find the unknown length.

b
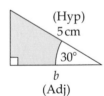
(Hyp)
5 cm
30°
b
(Adj)

Cos uses adjacent and hypotenuse.

$\cos 30° = \dfrac{\text{adj}}{\text{hyp}} = \dfrac{b}{5}$

$b = 5 \times \cos 30° = 4.3 \ (1 \text{dp})$

Exercise 14d

1 Use a calculator to find these values, rounding your answers to 3 significant figures.

   a sin 60°   **b** cos 45°   **c** tan 37.5°

   d sin 36°   **e** tan 48.5°   **f** cos 40.4°

   g sin 70.6°   **h** cos 20.3°   **i** tan 38.7°

2 Find the angle θ, correct to one decimal place, if

   a $\sin \theta = 0.5$ **b** $\cos \theta = 0.5$ **c** $\tan \theta = 1$

   d $\sin \theta = 0.75$ **e** $\cos \theta = 0.25$ **f** $\tan \theta = 1.5$

3 Calculate the angle θ in each right-angled triangle. Give your answers to one decimal place.

a

b

c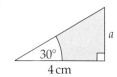

d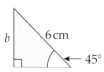

e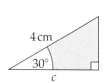

f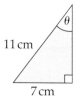

> ### Did you know?
>
> Using trigonometry, the height of the hill Mynydd Graig Goch in North Wales was calculated as 609 m (< 2000 feet). In 2008 it was re-measured using GPS to be 609.75 m. It is now officially a mountain (over 2000 feet)!

4 Calculate the unknown length in each triangle. Give your answers to one decimal place.

a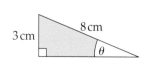

b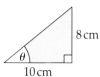

c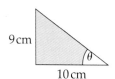

d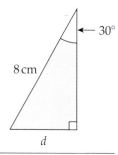

Problem solving

5 ABC is an isosceles triangle, so that AB = AC = 5 cm. Angle ABC is 30°. Calculate the area of triangle ABC.

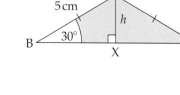

6 The sides of a regular pentagon are 10 cm long. Calculate the length of a diagonal.

〈 p.76

You use a **bearing** to give the direction of one point from another point.

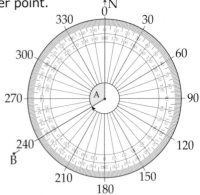

> ⬤ A **three-figure bearing** is the angle
> ▶ measured from north (N)
> ▶ in a clockwise direction
> ▶ and written with three digits

Examples are 047°, 147°

You can use angle properties to solve problems using bearings.
The bearing of A from B is 060°.

The bearing of B from A is 240°.

The bearing from B to A is called the **reverse bearing**, it is 240°.

Notice that 240° − 180° = 60°
and 60° + 180° = 240°

You either add or subtract 180° to find the reverse bearing.

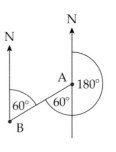

> The reverse bearing is sometimes called the **back bearing**.

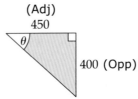

You can use trigonometry to solve problems using bearings.

Example

Julia walks north for 400 metres, then west for 450 metres. Calculate the bearing of the starting point from the finishing point.

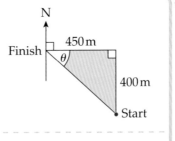

First calculate the angle θ.
Tan uses opposite and adjacent.

(Adj)
450

θ

400 (Opp)

$$\tan \theta = \frac{\text{opp}}{\text{adj}} = \frac{400}{450} = 0.888 \ldots$$
$$\theta = \tan^{-1} 0.888 \ldots = 41.6° \ (1\,dp)$$

The bearing of the Start from the Finish is $90° + 41.6° = 131.6°$

Exercise 14e

1 Calculate the bearing of **i** B from A **ii** A from B.

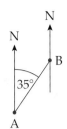 **a** 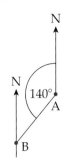 **b** 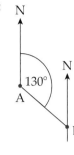 **c** 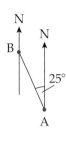 **d**

Problem solving

2 Calculate the bearing of
   a Belfast from Dublin
   b Sligo from Dublin
   c Sligo from Belfast
   d Dublin from Sligo
   e Belfast from Sligo.

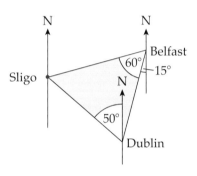

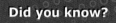
3 Lance runs west for 5 km
   and then south for 8 km.
   Calculate
   a the distance Lance is from his starting point
   b the bearing of the starting point from the finishing point.

4 A ship is due east of a buoy.
   It sails on a bearing of 210° for 50 km
   until it is due south of the buoy.
   How far is the ship from the buoy now?

5 A speedboat leaves the harbour on a bearing
   of 060° and travels 8 nautical miles.
   It then travels on a bearing of 150° for a
   further 6 nautical miles.
   a Draw a scale drawing to show the voyage.
   b Measure the distance and bearing of the
      speedboat from the harbour.
   c Use trigonometry to calculate the distance
      and bearing of the speedboat from the harbour.

Check out

You should now be able to ...

Test it ➡

		Questions
✓ Classify 3D shapes and draw 2D representations.	6	1
✓ Calculate the surface area and volume of a prism.	8	2
✓ Use Pythagoras' theorem in three dimensions.	8	3
✓ Use sine, cosine and tangent to find lengths and angles in right-angled triangles.	8	4 – 7
✓ Use trigonometry in calculations with bearings.	8	8

Language	Meaning	Example
Plane of symmetry	A plane that cuts a solid into two identical halves.	
Opposite side	The side in a right-angled triangle directly opposite the angle being considered.	
Adjacent side	The side in a right-angled triangle next to the angle being considered.	
Sine (sin)	The ratio $\dfrac{\text{opposite}}{\text{hypotenuse}}$	$\sin 30° = 0.5$
Cosine (cos)	The ratio $\dfrac{\text{adjacent}}{\text{hypotenuse}}$	$\cos 30° = 0.866$
Tangent (tan)	The ratio $\dfrac{\text{opposite}}{\text{adjacent}}$	$\tan 30° = 0.577$

1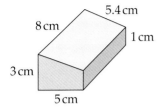

5.4 cm
8 cm
1 cm
3 cm
5 cm

   a How many faces, edges and vertices does the prism have?

   b Draw a net of the prism.

   c Draw the plan view of the prism.

2 Calculate

   a the length of AC

   b the volume

   c the surface area.

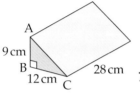

3

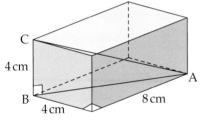

Calculate

   a the length AB   **b** the length AC

Write your answers in simplified surd form.

4 Calculate sin θ, cos θ and tan θ to 3 sf for this triangle.

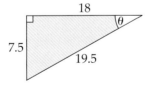

5 Find the angle θ, to 1 dp if

   a $\sin \theta = 0.6$

   b $\cos \theta = 0.2$

   c $\tan \theta = 1.4$

6 Calculate the angle θ to 1 dp in each triangle.

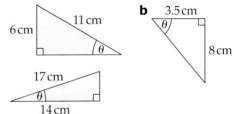

7 Calculate each unknown length x to 1 dp.

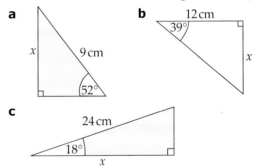

8 A ship is due west of a lighthouse. It sails on a bearing of 120° for 25 km at which point it is due south of the lighthouse.

   a How far is the ship from the lighthouse now?

   b On what bearing would the ship have to travel to return directly to its start point?

What next?

Score	0 – 3		Your knowledge of this topic is still developing. To improve look at Formative test: 3C-14; MyMaths: 1098, 1111, 1131, 1133 and 1086
	4 – 6		You are gaining a secure knowledge of this topic. To improve look at InvisiPen: 325, 327, 328, 374, 382, 383 and 384
	7 – 8		You have mastered this topic. Well done, you are ready to progress!

1 This cuboid has three planes of symmetry.

   a Draw a diagram to show each plane of symmetry.

   b Draw a net of the cuboid.

   c Draw accurate drawings of the front elevation, the side elevation and the plan view of the solid.

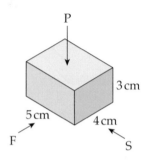

2 Calculate the surface areas and volumes of these prisms.

   a

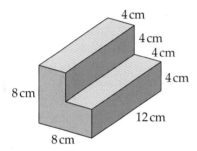

   b

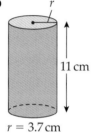

   c

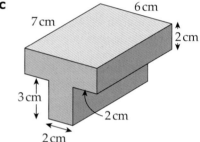

3 A cuboid measures 16 cm by 18 cm by 24 cm.

   Calculate

   a the length PQ

   b the length RP.

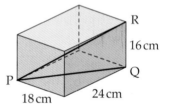

4 **a** Construct the right-angled triangle ABC.

   b Measure the length BC.

   c Calculate the value of $\frac{BC}{AC}$.

   d Use a calculator to find the value of tan 40°.

   e Explain why triangles PQR and XYZ are similar to triangle ABC.

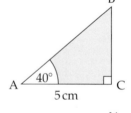

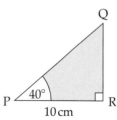

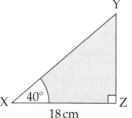

   f Calculate the lengths QR and YZ.

5 **a** If $\sin 30° = \frac{1}{2}$, what values could a and b take?

b Write an equation relating a to b.

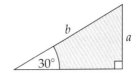

6 Calculate the angle θ in each right-angled triangle.
Give your answers to one decimal place.

a 20 cm 14 cm θ

b θ 12 cm 4 cm

c 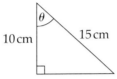 θ 10 cm 15 cm

d 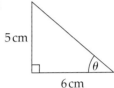 5 cm 6 cm θ

7 Calculate the unknown length in each triangle.
Give your answers to one decimal place.

a 10 cm 60° a

b 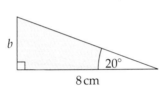 b 20° 8 cm

c 7.5 cm 45° c

d  25° 5 cm d

8 Toledo and Madrid are equidistant from Salamanca.
Calculate the bearing of

a Madrid from Toledo

b Salamanca from Toledo

c Toledo from Salamanca

d Madrid from Salamanca

e Toledo from Madrid.

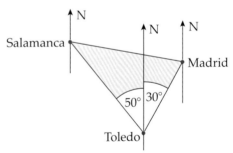

9 A ship is 40 km due north of a lighthouse.
It sails until it is 20 km due east of the lighthouse.
Use trigonometry to calculate

a the bearing the ship sailed on

b the distance the ship sailed.

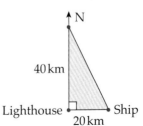

Case study 5: The golden rectangle

The golden rectangle has fascinated scholars for over 2000 years. It's a special kind of rectangle, which is often found in art and architecture.

Task 1

Rectangles come in all shapes and sizes, or different **proportions**. Here are six different rectangles. They can be sorted into three pairs of **similar** rectangles.

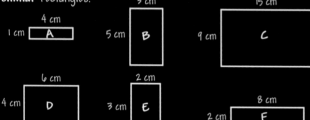

a Write down which pairs of rectangles are similar.
b For each rectangle, divide the longer side by the shorter side and write down the result.
  i What do you notice?
  ii What can you say about similar rectangles and the **ratio** of their sides?

Task 2

Here is a square with a smaller square next to it. Together they form a larger rectangle.

a Look at the larger rectangle.
  Divide the longer side by the shorter side and write down the result, to 1 d.p.
b Now look at the smaller rectangle and do the same. What do you notice? Describe your findings using the word 'similar' if possible.

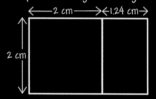

Task 3

Here is a **golden rectangle**.
The smaller rectangle is similar to the larger rectangle.
You can write this as a formula:

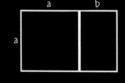

$$\frac{a + b}{a} = \frac{a}{b}$$

a For each rectangle, decide if it is golden or not. Use a calculator to help you, rounding your answers to 1 d.p. If there is a slight difference in your calculator answers, suggest why this might be the case.

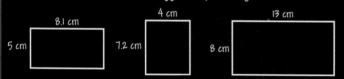

b (challenge) A golden rectangle has shorter length 3 cm.
  i See if you can find its longer length to 1 d.p.
  ii Construct the rectangle as accurately as you can.
c The number you get when you divide the longer side of a golden rectangle by the shorter side is called the **golden ratio**. Write down its value to 1 d.p.

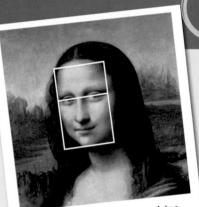

Leonardo da Vinci, Mona Lisa
The face fits within a golden rectangle. A smaller golden rectangle splits the face at the eye line.

Task 4

Look at the portrait of Mona Lisa, which is shown on this page.
By measuring lengths, describe why her face is framed by a golden rectangle.

Task 5

a How to CONSTRUCT A GOLDEN RECTANGLE

Draw a square.
Mark the mid point
of the base.

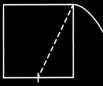

Set a pair of compasses to
the distance between the
mid point and the top
corner and draw an arc.

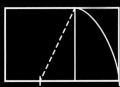

Extend the base of
the square to the arc
and complete the
rectangle.

b Without measuring, find the value of the golden ratio from this construction.

Task 6

In the **Fibonacci series**, each term is generated by adding the
previous two terms.

$$0, 1, 1, 2, 3, 5, 8, 13, \ldots$$

a Find the ratio of any two adjacent numbers from the series,
dividing the larger one by the smaller one.

b Try this for several pairs of adjacent numbers, working towards
the larger numbers.

What do you notice as you use larger and larger numbers?

The Fibonacci series can be shown as a set of squares.

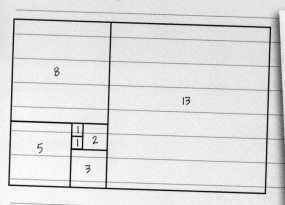

c Draw your own set of
squares in this way.

For each new size of rectangle
that you produce, find the ratio
of its length to its width.

d What do you notice about
the ratios?

What does this tell you
about the rectangle?

e You can draw arcs in each square
to produce a **Fibonacci spiral**.

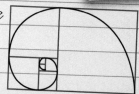

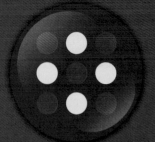

15 Ratio and proportion

Introduction

All of us need to eat! Eating provides the fuel that makes our bodies work but too much fuel can have serious consequences for our health. One way of burning off extra kilocalories is by walking. The faster you walk and the further you walk, the more kilocalories you will burn off. You can say that the number of kilocalories burned is proportional to the amount of exercise – or to put it another way: double the pain, double the gain!

What's the point?

Understanding proportion means that you can work out the effects of doubling and halving quantities.

Objectives

By the end of this chapter, you will have learned how to ...
- Describe proportion using fraction notation.
- Calculate fractional change.
- Solve problems involving ratio.
- Solve problems using direct proportion and scale factors.
- Interpret maps and scale drawings.
- Solve problems involving proportional reasoning, including financial problems.

Check in

1 Calculate the percentage change using a single multiplier.

    a   Increase £220 by 7.2%.   **b**   Decrease 45 kg by 3.75%.

2 **a**   A newborn baby weighs 2.91 kg. After 7 days the baby weighs 2.65 kg.
        Calculate the percentage reduction in the baby's weight.

    b   At age 3, Olivia is 68 cm tall. At age 5, she is 1 m tall.
        How many times taller has Olivia become?

3 Round each number to one significant figure.

    a   27.5   **b**   0.0772   **c**   2786   **d** 3.78×10^2 **e** 4.68×10^{-2}

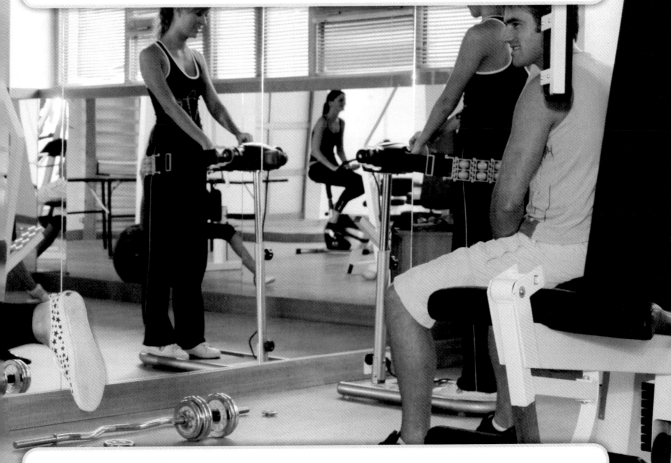

Starter problem

When you walk 3 miles you burn off 243 kilocalories.

How far do you have to walk to burn off the extra calories in a chocolate bar?

How far would you have to walk if you had to burn off the calories you eat at lunch today?

Investigate.

15a Fractions and proportion

Example

What proportion of the rectangle is shaded?

The area of the rectangle = 12 square units

$\quad$ Area of triangle = $\frac{1}{2}$ × base × height

$\quad\quad$ Area of A = $\frac{1}{2}$ × 2 × 3 = 3

$\quad\quad$ Area of B = $\frac{1}{2}$ × 2 × 2 = 2

$\quad$ The area of A + B = 5 square units

$\quad$ Proportion shaded = $\frac{5}{12}$

fraction = $\dfrac{\text{Area shaded}}{\text{Total area}}$

Write the proportion as a

⬤ You can find a fraction of a number or quantity using multiplication.

Example

Calculate $\frac{7}{15}$ of 40 km.

$\frac{7}{15}$ of 40 = $\frac{7}{{}^3 15}$ × ${}^8 40$ = $\frac{56}{3}$ = $18\frac{2}{3}$ km

Give your final answer as a mixed number.

⬤ You can calculate a fractional increase or decrease in a single calculation using multiplication.

Example

a Decrease £235 by $\frac{1}{5}$

b Increase £235 by $\frac{1}{5}$

a The price has decreased by $\frac{1}{5}$

$\quad$ New price

$\quad = (1 - \frac{1}{5})$ of the old price

$\quad = \frac{4}{5}$ of the old price

$\quad = \frac{4}{5}$ × 235

$\quad = £188$

b The price has increased by $\frac{1}{5}$

$\quad$ New price

$\quad = (1 + \frac{1}{5})$ of the old price

$\quad = \frac{6}{5}$ of the old price

$\quad = \frac{6}{5}$ × 235

$\quad = £282$

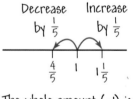

Decrease $\quad$ Increase

by $\frac{1}{5}$ $\quad\quad$ by $\frac{1}{5}$

$\frac{4}{5}$ $\quad$ 1 $\quad$ $1\frac{1}{5}$

The whole amount (=1) is increased or decreased by $\frac{1}{5}$.

Re-write $1\frac{1}{5}$ as an improper fraction $\frac{6}{5}$.

Exercise 15a

1 Calculate the following, leaving your answers as mixed numbers where appropriate.

   a $\frac{5}{9}$ of 12 feet

   b $\frac{8}{15}$ of 50 m

   c $\frac{3}{7}$ of 21 m

   d $\frac{7}{12}$ of 42 sec

   e $1\frac{7}{16}$ of 44 km

   f $2\frac{3}{5}$ of 8 miles

   g $1\frac{4}{9}$ of 75 g

   h $3\frac{8}{15}$ of $\frac{3}{4}$ hr

2 Calculate these fractional changes, leaving your answer as a mixed number where appropriate.

   a Increase £40 by $\frac{2}{5}$

   b Decrease £20 by $\frac{3}{5}$

   c Increase 12 kg by $\frac{1}{5}$

   d Decrease 50 kg by $\frac{1}{3}$

   e Increase 25 miles by $\frac{2}{7}$

   f Decrease 120 miles by $\frac{2}{9}$

   g Increase $3\frac{1}{2}$ km by $\frac{1}{4}$

   h Decrease $45\frac{1}{4}$ kg by $\frac{1}{5}$

Problem solving

3 Express the shaded shape in each diagram as a fraction of the square.

   a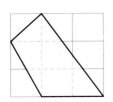

   b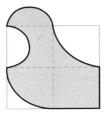

4 Here are two shapes.

   a What fraction of shape B is shape A?

   b What fraction of shape A is shape B?

Shape A   Shape B

5 a Gina weighs 64 kg. Twelve months later her weight has increased by $\frac{1}{12}$. What is Gina's new weight?

   b In a sale all prices are reduced by $\frac{1}{3}$. A CD costs £14 before the sale. What is the sale price of the CD?

6 Boris buys a company for £2 million pounds. Each year he wants to increase the size of his company by $\frac{1}{5}$.
How big will the company be in

   a two years time

   b five years time

   c ten years time?

   d Investigate by how much the company will grow for different numbers of years.

Did you know?

Thomas Malthus was an economist and demographer. He noticed that populations tend to grow by a constant factor, 1, 2, 4, 8, 16,... whilst food supplies only grow by a constant amount, 1, 2, 3, 4, 5,... So, usually, food supply restricts population growth.

MyMaths.co.uk

15b Ratio and proportion

You can solve problems using **ratios**.

If you and a friend share a pie in the ratio $1:3$ then you get $\frac{1}{4}$ of the pie!

Example

Brass is made from zinc and copper.
Johann mixes 3 kg of zinc with 4.5 kg of copper.
  a How much copper does Johann need to mix with 25 kg of zinc?
  b How much zinc does Johann need to mix with 50 kg of copper?

Simplify the ratio.

zinc : copper

$3:4.5$

$\times 2 \Big(\quad \Big) \times 2$

$6:9$

$\div 3 \Big(\quad \Big) \div 3$

$2:3$

You simplify a ratio by dividing or multiplying all parts of the ratio by the same number.

zinc : copper $= 1:\frac{3}{2} = \frac{2}{3}:1$

  a Amount of copper $= \frac{3}{2} \times$ amount of zinc

$= \frac{3}{2} \times 25\,\text{kg} = 37.5\,\text{kg}$

$\times \frac{3}{2}$

zinc : copper

$\times \frac{2}{3}$

The ratio tells you how many times bigger each part is compared to the other part.

  b Amount of zinc $= \frac{2}{3} \times$ amount of copper

$= \frac{2}{3} \times 50\,\text{kg} = 33\frac{1}{3}\,\text{kg}$

You can compare ratios by expressing them in the form $1:n$.

Example

Simeon is investigating the ratio of potato to other
ingredients in two pies. The ratio is $7:20$ for cottage pie
and $9:25$ for shepherd's pie.
Which pie contains more potato as a proportion?

The ratio tells you how many times more other ingredients there are compared to potato.

Express both ratios in the form $1:n$

<u>Cottage pie</u>
potato : other

$\div 7 \Big(\quad 7:20 \quad \Big) \div 7$

$1:2.86$ (2 dp)

<u>Shepherd's pie</u>
potato : other

$\div 9 \Big(\quad 9:25 \quad \Big) \div 9$

$1:2.78$ (2 dp)

Shepherd's pie has the greater proportion of potato.

Exercise 15b

1 Write each ratio in its simplest form.

 a $5:10$ **b** $10:5$

 c $8:2$ **d** $84:12$

 e $3:9:15$ **f** $100:25:125$

 g $\frac{1}{2}:\frac{1}{4}$ **h** $\frac{1}{3}:\frac{5}{6}$

 i $75\%:25\%$ **j** $0.4:3$

 k $0.5:2$ **l** $0.\dot{2}:0.\dot{9}$

 m $1.6:2.4$ **n** $2.1:3.5:4.2$

 o $2:2.4:3$ **p** $1.2\,\text{m}:200\,\text{cm}$

 q $4.5\,\text{kg}:2000\,\text{g}$ **r** £$3.60:80\text{p}$

2 Express each of these ratios in the form $1:n$.

 a $2:8$ **b** $4:10$

 c $3:20$ **d** $5:75$

 e $4:11$ **f** $3:22$

 g $22:3$ **h** $11:4$

 i $\frac{1}{4}:\frac{1}{2}$ **j** $0.3:0.9$

 k $400:100$ **l** $100:5$

 m $500\,\text{g}:750\,\text{g}$ **n** $4t:\frac{1}{2}t$

 o $1\,\text{hr}:20\,\text{min}$ **p** $1\,\text{min}:20\,\text{sec}$

Problem solving

3 **a** The ratio of nylon to other materials in two T-shirts are $4:7$ and $9:16$.
   Which T-shirt has the greater proportion of nylon?

  b The ratio of teachers to students in Appleby School is $3:40$, compared to
   Plumley Academy where the ratio is $4:51$.
   Which school has the higher proportion of teachers?

4 **a** John and Kaseem divide £21 in the ratio $3:7$.
   How much money does Kaseem receive?
   What proportion of the money does John receive?

  b A wall covers $70\,\text{m}^2$. The wall is painted blue, white and yellow in the ratio $5:4:5$.
   What area of the wall is painted blue? What proportion of the wall is painted yellow?

  c In a sports club the ratio of men to women is $7:9$.
   i If there are 272 members, how many are women?
   ii If there are 91 men, how many women are there?

5 **a** Sarah is 1.6 m tall. Keenan is 15% taller than Sarah.
   What is the ratio of Sarah's height to Keenan's height?

  b Orange paint is made using 2 parts of red paint mixed with 3 parts of
   yellow paint. Aleshia has 450 ml of red paint and 700 ml of yellow paint.
   What is the maximum volume of orange paint she can make?

6 The ratio of area C to area B is $1:2$.
 Find the area of A.

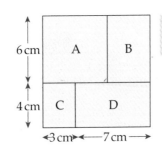

The diagram is not drawn to scale.

> When two quantities are in **direct proportion** the ratio between the quantities is fixed.

Example

The table shows the number of inches and the corresponding number of centimetres.

Inches	3	5	6	10	15	50
Centimetres	7.5	12.5	15	25	37.7	125

a What is the ratio of inches to centimetres?

b Are inches and centimetres in direct proportion?

a Inches : centimetres

$3 : 7.5$

$6 : 15$

$2 : 5$

$1 : 2.5$

$\times 2.5$

$\times \frac{5}{2}$

inches : centimetres

2 : 5

$\times \frac{2}{5}$

$\times 0.4$

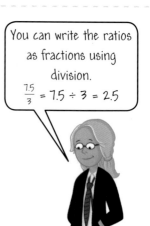

You can write the ratios as fractions using division.

$\frac{7.5}{3} = 7.5 \div 3 = 2.5$

b Inches and centimetres are directly proportional because all the ratios are equal.

$$\frac{\text{Centimetres}}{\text{Inches}} = \frac{7.5}{3} = \frac{12.5}{5} = \frac{15}{6} = \frac{25}{10} = 2.5$$

Number of centimetres = 2.5 × number of inches

> You can use fractions to solve direct proportion problems.

Example

Brown paint is made by mixing 7 parts of red paint with 4 parts of green paint. How many litres of red paint should you mix with 23 litres of green paint?

Express the ratios in fractional form.

$$\frac{\text{Red}}{\text{Green}} = \frac{7}{4} = \frac{\square}{23}$$

Amount of red $= \frac{7 \times 23}{4}$

$= \frac{161}{4}$

$= 40\frac{1}{4}$ litres

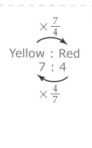

$\times \frac{7}{4}$

Yellow : Red

7 : 4

$\times \frac{4}{7}$

red $= \frac{7}{4} \times$ green

green $= \frac{4}{7} \times$ red

Exercise 15c

1 Identify which of these sets of data are in direct proportion.
Explain and justify your answers.

a

Weight (kg)	2	3	6	10	15	50
Cost (£)	2.50	3.75	7.50	12.50	18.75	62.50

b

x	1	3	6	10	15	50
y	6.5	19	37	61	90	275

c

x	3	5	7	9	11	13
y	3.6	6	8.4	10.8	13.2	15.6

Problem solving

2 Solve each of these problems.

a 3 kg of carrots cost 75p. What is the cost of 8 kg of carrots?

b A recipe for four people uses 600 g of potato. How much potato do you need for seven people?

3 Solve each of these problems.

a To make green paint, 6 parts of yellow paint are mixed with 5 parts of blue paint. How many litres of blue paint should you mix with 40 litres of yellow paint?

b Travis knows that 8 kilometres is the same distance as 5 miles. How far in miles is 25 km?

c Jackson changed £350 into €385. At the same exchange rate how many pounds will Jackson get for €200?

d In a sale all the prices in the shop are reduced by the same percentage. A coat is reduced in price from £95 to £66.50. A pair of shoes normally costs £108. What is the sale price of the shoes?

e A model of a plane is 1.2 m long. The plane is 9 m long in real life. If the wingspan of the real plane is 13 m, what is the wingspan of the model?

> Try to use the fractional method.

> **Did you know?**
>
>
>
> Suprabha Beckjord is the only woman to have finished the self-transcendence 3100 mile race, the world's longest certified footrace. Her fastest time is 49 days 14.5 hours.

4 Gina reads a newspaper that says a man can run 1 km in 3 minutes. She works out that he can run 40 km in 2 hours. Explain why this might not be correct.

5 Gavin draws some circles with different radii. For each one he works out the area and records his answers in a spreadsheet.

‹ p.22

	A	B	C	D	E	F
1	Radius	2	3	5	10	20
2	Area	12.572	28.27	78.54	314.16	1256.64

Investigate to see if the area of a circle is directly proportional to the radius of the circle.

15d Proportion and scale

‹ p.162

⬤ When a photograph or shape is enlarged by a **scale factor**, the ratio of the length and width stays the same.

Example

Calculate the length of the larger photograph.

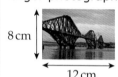

8 cm

12 cm

14 cm

☐

$$\frac{length}{width} = \frac{12}{8} = \frac{\square}{14} = 1.5$$

or

$$\frac{enlarged\ width}{original\ width} = \frac{14}{8} = \frac{\square}{12} = 1.75$$

length = 1.5 × width

scale factor = 1.75

enlarged length = $\frac{12}{8}$ × 14

enlarged length = $\frac{14}{8}$ × 12

= 1.5 × 14

= 1.75 × 12

= 21 cm

= 21 cm

The ratio of length and width stays the same.

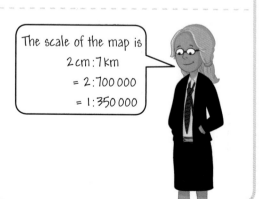

⬤ You can use direct proportion to solve problems involving maps and scale drawings.

Example

A map is drawn on a scale of 2 cm to represent 7 km.
The real distance between two towns on the map is 30 km.

What is the distance on the map between the towns?

Map scale = $\frac{Map\ measurement}{Real\ measurement}$

= $\frac{2\,cm}{7\,km} = \frac{\square}{30\,km}$

Distance on map = $\frac{2\,cm \times 30\,km}{7\,km}$

= $\frac{60\,cm}{7}$

= $8\frac{4}{7}\,cm$

= 8.6 cm (1 dp)

The scale of the map is
2 cm : 7 km
= 2 : 700 000
= 1 : 350 000

6

Exercise 15d

1 Write the following scales in the form $1:n$
- **a** 1 cm represents 1 km
- **b** 2 cm represents 5 km
- **c** 3 cm represents 27 m
- **d** 20 cm represents 45 m
- **e** 7 mm represents 42 cm
- **f** 6 mm represents 60 cm

2 What does 5 cm represent on a map with these scales?
- **a** 1:2000
- **b** 1:25000
- **c** 1:50000
- **d** 1:12
- **e** 1:100000
- **f** 1:4500
- **g** 1:350
- **h** 1:100
- **i** 1:3000000
- **j** 1:1600000

Problem solving

3 Each pair of photographs shows the original photo and its enlargement. Calculate the missing length in each pair.

Give answers to 1 dp

a

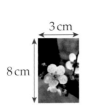

3 cm
8 cm
☐
12 cm

b

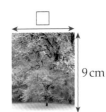

12 cm
15 cm
☐
9 cm

4 a A map has a scale of 1:25000.
- **i** What distance does 3.8 cm on the map represent in real life?
- **ii** What distance on the map represents a real-life measurement of 8 km?

b A model of a plane is built to a scale of 1:24. The wingspan of the real plane is 12.8 m. How long is the wingspan of the model?

5 A map scale uses 3 cm to represent 20 km.
The real distance between two towns is 13 km.
- **a** What is the distance on the map between the towns?
- **b** Write the scale of the map as a ratio in the form $1:n$.

6 Mohinder draws a rectangle and works out its area.
He enlarges his rectangle by different scale factors and he records the areas of the new rectangles in a spreadsheet.

	Original	Enlargement A	Enlargement B	Enlargement C	Enlargement D	Enlargement E
Lenght	6 cm	12 cm	18 cm	9 cm	4.8 cm	7.5 cm
Width	4 cm	8 cm	12 cm	6 cm	3.2 cm	
Area						

- **a** Copy and complete the spreadsheet.
- **b** Write the scale factor Mohinder used for each enlargement.
- **c** Investigate how the area of the rectangle changes after each enlargement.

> When two quantities are in **direct proportion** you can use the **ratio** to change from one quantity to another using a single **multiplier**.

Example

A picture is enlarged by a scale factor of 1.5. What is the ratio of lengths in the original picture to corresponding lengths in the enlargement?

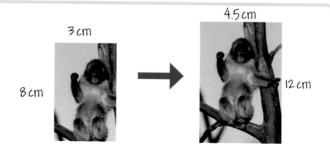

original : enlarged

$= 3:4.5 = 8:12$

$= 2:3$

or

$= 1:$ scale factor

$= 1:1.5$

$= 1:\dfrac{3}{2}$

$= 2:3$

> You can also use the ratios between lengths to solve direct proportion problems.

Example

A picture measures 8 cm by 3 cm. The picture is enlarged so that the new picture has a length of 12 cm. What is the width of the enlarged picture?

$\dfrac{\text{width}}{\text{length}} = \dfrac{3}{8} = \dfrac{\square}{12}$

Enlarged width $= \dfrac{3}{8} \times 12$

$= 4.5$ cm

width $= \dfrac{3}{8} \times$ length

length $= \dfrac{8}{3} \times$ width

7

Exercise 15e

1 Write

  i each pair of lengths as a ratio

  ii write the fraction multiplier between them.

  a 1 cm, 4 cm   **b** 2 m, 3 m   **c** 4 m, 24 m   **d** 3 cm, 7.5 cm

  e 20 cm, 25 cm   **f** 8 m, 5 m   **g** 3 m, 60 cm   **h** 7 km, 4 km

Problem solving

2 Solve these problems.

  a Joe mixes 7 parts of black paint with 4 parts of white paint to make grey paint. If he has 30 litres of black paint, how many litres of white paint does he need to make grey paint?

  b In a sale all the prices in the shop are reduced by the same percentage. A pair of trainers is reduced in price from £120 to £79.20. A pair of football boots cost £49.50 in the sale. What was the price of the football boots before the sale?

3 A triangle is enlarged by a scale factor.

  a Calculate the ratio of lengths in the original triangle to those in the enlarged triangle.

  b Calculate the lengths of the sides marked x and y.

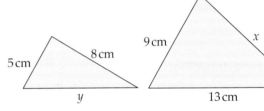

4 A ship sails across the Pacific Ocean. It has an average speed of 31 miles per hour. It uses fuel at the rate of 1 litre every 2.4 feet travelled. Calculate to the nearest litre how much fuel the ship uses to travel

  a for 1 hour   **b** in a journey of 4300 km.

1 mile = 5280 feet
       = 1.609 km

5 Henry fills a glass with blackcurrant cordial and water in the ratio 1 : 4. He drinks $\frac{1}{4}$ of the liquid and then fills the glass with more water. What is the ratio of blackcurrant cordial to water in the glass now?

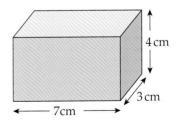

6 Mandy makes this box.

  a Calculate the volume and surface area of the box.

  She makes a second box 1.5 times larger than the first.

  b Calculate the volume and surface area of the new box.

  c How have the surface area and volume changed after the enlargement?

  d Investigate enlarging the original box by different scale factors and in each case note how the surface area and volume have changed.

  e Mandy makes a third box which is an enlargement of the original box. The volume of this box is 1312.5 cm³. What are the dimensions of the new box?

The length, width and height are all 1.5 times longer.

Nicole is a student taking a photography course at university.

Example

Nicole wants to buy a brand new camera, priced at £475. She already has £300 **saved**. She can save £30 per month. After how many months would Nicole have saved up enough money for the camera?

475 − 300 = 175

175 ÷ 30 = 5.833333333   Use a calculator

It will take Nicole 6 months.   Round up

Nicole needs to **budget** carefully as she has limited money to live on.

Example

Nicole has £900 to live on each month. She plans her monthly **outgoings** on a spreadsheet.

Outgoings	Amount budgeted (£)
Accommodation	400
Food shopping	180
Household bills	50
Transport	50
Mobile phone	20
Clothes/other	30
Entertainment	80

a How much money does Nicole have left over each month?

b What proportion of Nicole's monthly expenses goes on
  i   accommodation
  ii food shopping?
In each case, give your answer as a fraction in its lowest terms.

Budget Plan

The left-over amount is often referred to as **contingency**, which can be used for unforeseen expenses.

a   total amount = £810

    900 − 810 = 90   Nicole has £90 left over.

b i $\frac{400}{900} = \frac{4}{9}$ ii $\frac{180}{900} = \frac{1}{5}$

When Nicole goes to the supermarket, she makes choices based on **value for money**.

Example

A particular brand of orange juice comes in three different sizes. Nicole wants to choose the size that gives her the best value for money. Help Nicole make her choice.

To compare value, calculate the price for 1 litre (**unitary method**).

Size X 0.99 × 4 = £3.96   4 lots of 250 ml in 1 litre.

Size Y 1.80 × 2 = £3.60   2 lots of 500 ml in 1 litre.

Size Z 2.50 × $\frac{4}{3}$ = £3.33   ÷ 3 to find cost of 250 ml

                    then ×4 to scale up to 1 litre.

Size Z is the cheapest per litre, so would give the best value for money.

£0.99 for 250 ml
Size X

£1.80 for 500 ml
Size Y

£2.50 for 750 ml
Size Z

Exercise 15f

1 Work out how long it would take to be able to afford each of these items, starting from nothing and at the rate of saving given.

   a 32 GB tablet, price £320, can save £50 per month

   b 18 inch laptop, price £299, can save £40 per month

   c Classic designer boots, price £160, can save £45 per month

   d 100 watt guitar amp, price £429.95, can save £28 per week.

2 Laura has created a budget for her monthly expenses.

   a Laura takes home £1200 **salary** each month. How much does she have left over each month for herself?

   b Work out the proportion of Laura's salary for each of the six categories in the spreadsheet. In each case, give your answer as a fraction in its lowest terms.

   c Draw a pie chart of Laura's monthly spending. Label your diagram clearly.

Expense type	Amount budgeted £
Accommodation	500
Food shopping	200
Household bills	50
Transport	60
Other essential expenses	90
Amount left over	

   d Laura's friend Marie says '£1200 per month means you take home £400 per week.' Describe why this is only an *estimate*, and explain whether Laura's weekly pay will be more or less than this.

Problem solving

3 Oscar is deciding which of these two brands of coffee to buy, *Heavenly Blend, Coffee Club* or *Mountain Brand*. He thinks: 'I like all of them. I can afford to buy any of them. I just want good value for money.' Which one would you advise Oscar *not* to buy? Give your reasoning.

£4.99 for 250g   £3.99 for 200g   £2.25 for 100g

4 Maisie likes houmous! Two competing brands are of similar quality, but they now both have offers. Which should Maisie choose, and why?

5 Look again at Nicole's budget spreadsheet on the opposite page. Due to **price inflation**, the amount she pays for accommodation, food shopping and household bills have all gone up by 12%. She still has only £900 to spend each month. Advise Nicole on where she can make a saving to ensure that she still has £90 left over as contingency.

Brand X normally £1.50 for 200g, now 50% off!   Brand Y normally £1.50 for 200g, now 50% extra free!

Check out

You should now be able to ...

Test it ➡

Questions

✓ Describe proportion using fraction notation.	6	1
✓ Calculate fractional change.	6	2, 3
✓ Solve problems involving ratio.	7	4 – 7
✓ Solve problems using direct proportion and scale factors.	7	8, 9
✓ Interpret maps and scale drawings.	7	10
✓ Solve problems involving proportional reasoning, including financial problems.	7	11, 12

Language	Meaning	Example
Proportion	When change in one variable causes change in another variable (or variables).	If one bucket costs £3 then 5 buckets will cost £15
Repeated percentage	Applying a percentage change more than once.	£20 increased by 5% per year for 2 years is £22.05
Decimal multiplier	A single decimal used to calculate a percentage change.	Decrease by 15% $\times$ 0.85
Ratio	A way of expressing a proportion.	$2:5$
Direct proportion	When change in one variable causes a linear change in another variable.	If one bucket costs £3 then 5 buckets will cost £15
Scale factor	The ratio length of image : length of object.	A scale factor of 3 means all the lengths in the image are 3 times those of the object

1 Calculate the following leaving your answers as mixed numbers.

    a $\frac{3}{8}$ of £18

    b $10\frac{5}{9}$ of $\frac{3}{5}$ miles

2 Increase

    a £15 by $\frac{4}{9}$ **b** 840 kg by $\frac{1}{7}$

3 Decrease

    a 33 km by $\frac{2}{11}$ **b** $81\frac{1}{2}$ m by $\frac{2}{3}$

4 Write each ratio in its simplest form.

    a 3.2 : 1.6 : 4

    b 6.5 kg : 4000 g

5 Write each ratio in the form 1 : n

    a 8 : 28

    b 60 cm : 2.25 m

6 The ratio of red to blue in two shades of paint are 3 : 7 and 5 : 11. Which paint has the greater proportion of blue?

7 The ratio of lifeguards, adults and children in a swimming pool is 1 : 10 : 15

    a If there are 364 people in the swimming pool, how many are children?

    b How many adults will there be if there are 117 children?

    c How many lifeguards are required if there are 200 other people in the pool?

8 250g of butter costs £2.10. What is the cost of 175 g of the same butter?

9 In a sale all items are reduced by the same percentage. A book is reduced from £19 to £12.35. A pen originally cost £9. What is its sale price?

10 A map has a scale of 1 : 35 000

    a What distance does 3 cm on the map represent in real life?

    b What length on the map represents a real-life measurements of 10.5 km?

11 A car uses fuel at the rate of 1 litre every 14 km when travelling at an average of 50 km/h.
Calculate to the nearest millilitre how much fuel the car uses to travel

    a 100 km

    b for 45 minutes.

12 Three bottles of the same brand of olive oil are on sale:

> bottle A holds 250 ml and is priced at £1.99

> bottle B holds £500 ml and costs £3.49

> bottle C holds 750 ml and costs £5.99

Which bottle represents the best value for money?

What next?

Score			
	0 – 5		Your knowledge of this topic is still developing. To improve look at Formative test: 3C-15; MyMaths: 1073, 1052, 1037, 1103, 1037 and 1245
	6 – 9		You are gaining a secure knowledge of this topic. To improve look at InvisiPen: 142, 155, 193, 195 and 196
	10 – 12		You have mastered this topic. Well done, you are ready to progress!

MyMaths.co.uk

1   Calculate these quantities, leaving your answers as mixed numbers where appropriate.

   a $\dfrac{3}{5}$ of 45 m   **b** $\dfrac{7}{15}$ of 20 km   **c** $1\dfrac{3}{7}$ of £42   **d** $2\dfrac{5}{8}$ of 30 kg

2   Calculate these fractional changes, leaving your answers as mixed numbers where appropriate.

   a Increase £80 by $\dfrac{7}{16}$ **b** Decrease £50 by $\dfrac{3}{10}$

3   At a school the ratio of boys to girls is 4 : 5.
There are 572 boys at the school.
How many children are there at the school?

4   A model of a yacht is built to a scale of 1 : 24.
The length of the yacht is 15.6 m.
What is the length of the model yacht?

5   The ratio of Lycra to other materials in two gym suits are 3 : 8 and 5 : 13.
Which gym suit has the greater proportion of Lycra?

6   **a** Sebastian and Belle divide £38 in the ratio 3 : 5.
    i How much money does Sebastian receive?
    ii What proportion of the money does Belle receive?
   b In a running club the ratio of men to women is 5 : 4.
    There are 196 women members.
    How many men are there in the club?

7   Solve these problems.
   a 5 kg of apples cost £1.85.
    What is the cost of 11 kg of apples?
   b A recipe for 6 people needs 500 ml of stock.
    How much stock do you need for 9 people?

8   Solve these problems.
   a To make pink paint mix 5 parts of white paint with 7 parts of red paint.
    How many litres of red paint should you mix with 12 litres of white paint?
   b Jenny changed £480 into $744 when she went to America.
    On her return she had $310 left.
    If the exchange rate stayed the same, how many pounds did she get back?

9 The distance on a map between Kerso and Larroz is 4.5 cm. In real-life the distance is 1.35 km.

   a Write the scale of the map as a ratio in the form $1:n$.

   b The distance on the map between Larroz and Horten is 18.2 cm. How far apart are the towns in real life?

10 A piece of A4 paper measures 210 mm by 297 mm. A piece of A3 paper is in proportion and has a length of 420 mm. What is the width of A3 paper? Write what you notice.

11 The larger triangle is an enlargement of the smaller triangle in each of these diagrams. Give your answers to 1 dp.

   a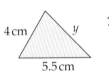

   b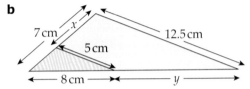

For each pair of triangles

   i calculate the ratio of the lengths in the original triangle to the lengths in the enlarged triangle

   ii calculate the lengths of the sides marked x and y.

12 Mariane's monthly expenses are as shown.

Expense type	Amount budgeted, £
Accommodation	600
Food shopping	300
Bills	80
Transport	100
Other essentials	120
Amount left over	300

   a Find the proportion of each expense type against the total.
Give your answers as a fraction in its lowest term.

   b Draw a pie chart of Mariane's expenses.

16 Probability

Introduction

Buying a new car or TV is a major purchase for a family. It is important when you spend a lot of money on an item that, it is reliable, works well and does not break down. However no product is 100% perfect. Most consumer items have experiments carried out on them to find out their probability of breaking or becoming defective under different conditions.

What's the point?

Probability theory is vitally important in ensuring the safety levels of cars, planes and trains.

Objectives

By the end of this chapter, you will have learned how to …

● Evaluate uncertainty and risk in real situations.
● Identify and calculate probabilities for independent events.
● Use tree diagrams to calculate probabilities.
● Calculate probabilities from experimental data.
● Use random numbers to model a situation.
● Calculate probabilities using a venn diagram.

Check in

1   An ordinary dice is rolled.

What is the probability of the score being

    a   a square number   **b**   not a square number?

2   Calculate

    a $\dfrac{1}{2} + \dfrac{1}{4}$ **b** $\dfrac{1}{2} \times \dfrac{3}{4}$ **c** $\dfrac{3}{5} + \dfrac{4}{10}$ **d** $\dfrac{2}{7} \times \dfrac{3}{8}$

    e $\dfrac{7}{10} \times \dfrac{4}{9}$ **f** $\dfrac{5}{4} + \dfrac{4}{9}$ **g** $\dfrac{1}{3} \times \dfrac{1}{2} + \dfrac{2}{3} \times \dfrac{1}{4}$ **h** $\dfrac{1}{5} \times \dfrac{3}{4} + \dfrac{2}{5} \times \dfrac{1}{2}$

Starter problem

This is a famous maths problem called the Monty Hall problem. It was part of an American game show.

There are three boxes.

One box has a real car inside, the others have model cars inside.

A contestant picks a box but does not open the box.

The host opens one of the other boxes and reveals a model car.

The host offers a switch. His unopened box for your unopened box.

Would you take it?

Investigate.

16a Prediction and uncertainty

How we deal with **risk** depends on a combination of how likely something is to happen and the consequences of it happening.

Probability is important because it helps you to manage uncertainty.

For each of the following, describe how big a problem it would be if you were delayed getting there when you planned to.

a   Meeting up with friends.

b   A job interview.

c   A flight to Spain for a holiday.

a   Not a big deal.

b   Being late would create a bad impression making it less likely to be offered a job.

c   Missing a flight is likely to mean missing the whole holiday or paying a lot of money to rebook onto another flight.

You can reduce the risk by setting out earlier and reducing the probability of being late.

Yulia wants to buy a car. She sees a car she likes in a garage for £3500. She sees the same car in a local newspaper as a private sale for £2950.

If she buys privately, a friend recommends getting a vehicle inspection report, costing £135.

Discuss any advantages and disadvantages you can see for buying each car and for the vehicle inspection report.

The garage is more expensive but Yulia will have a guarantee if there are problems with the car after she has bought it.

Buying a car privately, she has no guarantee, but it is cheaper. The vehicle inspection report would find anything major that is wrong and so is probably a good thing to get.

Exercise 16a

Problem solving

1 For each of the following, describe how big a problem it would be if you were delayed getting there when you planned to.
   a Getting to school on a normal school day.
   b Getting to school on a day when you have a GCSE.
   c Going to the cinema.

2 You are going out with friends to the park on a reasonably warm day but it looks as though it might rain.
   What are the reasons for and against taking a coat with you?
   Are there any other pieces of information which might make a difference to your decision?

3 You buy a new mobile phone costing £110.
   The cost of insuring the phone for 12 months is £34.99.
   a What are the advantages and disadvantages of taking the insurance?
   b If a group of 10 friends each took the insurance, how many would have to claim for a replacement phone for the group to save money overall? How likely do you think that is to happen?

4 Two friends have each been offered the chance to invest £100 000 in an oil field. A survey estimates a 75% chance that there is a substantial oil deposit, in which case they would get back double their money in a year's time. However, if there is no oil then they will lose their money.
   Mr Goldfinger is a millionaire businessman with enough ready cash to make the investment.
   Mrs Median has worked hard all her life in an office and has built up savings of £50 000 and owns her own home. She could take out a mortgage against her house to raise the other £50 000.
   Discuss what difference their financial position might make to whether or not they decide to invest and what you would advise each of them to do.

16b Independent events

You can apply probability to more than one event.

> ● If the outcome of one event does not affect what happens in another then the events are **independent**.

State, with reasons, whether these pairs of events are independent or not.

a Toss two coins and obtain

  i heads on the first coin   **ii** tails on the second coin.

b Roll a dice and obtain

  i a prime number   **ii** an even number.

c At a rugby game

  i it rains   **ii** the home side wins.

d You interview a final year student at university

  i they live at home   **ii** they have health problems.

a Yes.   What happens on one toss does not affect the second.

b No.   The primes are 2, 3 and 5. If you know there is a prime then only one out of three is even; without this information the chance is three out of six.

c Difficult to say.   The weather may change the likelihoods of the possible results but you would need to know more about the two teams to know how.

d Unlikely to be independent.   Students living at home are more likely to have a regular routine and be less likely to have health problems.

P(even given a prime) = $\frac{1}{3}$ whilst P(even) = $\frac{1}{2}$

> ● If A and B are independent events then P(A and B both happen) = P(A) × P(B)

A road safety officer estimates that $\frac{1}{4}$ of bicycles used by teenagers do not have properly inflated tyres and $\frac{1}{6}$ of them don't have adequate lights.

a If these are independent, what is the probability that a bicycle belonging to a teenager, chosen at random, has both faults?

b Do you think independence is reasonable?

Independence may be assumed even when not strictly true, to allow you to calculate an approximate answer.

a P(both faults) = $\frac{1}{4} \times \frac{1}{6} = \frac{1}{24}$.

b While the tyres and lights are not directly connected, they may both be dependent on how careful the owner is, so independence is not likely to be completely correct.

8

Exercise 16b

Problem solving

1. State, with reasons, whether these pairs of events are independent or not.
   a. A person visiting their doctor
      i. suffers from breathlessness
      ii. is a smoker.
   b. Roll a dice and obtain
      i. a square number
      ii. an even number.

2. For the second example, what is the probability that a bicycle belonging to a teenager, chosen at random, has neither fault?

3. The probability of throwing more than a 4 on a fair dice is $\frac{1}{3}$. What is the probability you will get over 4 both times if you throw it twice?

4. The probability that a person chosen at random in a company is married is 0.6. One third of the company employees are female.
   a. Assuming that these are independent, what is the probability that a person chosen at random in a company is a married female?
   b. Give reasons why the proportions of women and men in the company who are married might be different.

5. Discuss whether you think these pairs of events are independent or not.
   a. Roll a dice and obtain
      i. an odd number
      ii. an even number.
   b. Roll a red and a blue dice and obtain
      i. an even number on the red dice
      ii. an even number for the sum.
   c. You toss a coin four times and obtain
      i. a head on the first toss
      ii. you get the first tail on the third toss.

Did you know?

On an aeroplane, the designers try to have systems that work independently. So, for example, if a problem affects one engine, it can continue to fly on the other engines.

> These are all cases you could test experimentally. Does the proportion of times you see both events roughly equal the proportions of times you see each one multiplied together?

16c Tree diagrams

○ A **tree diagram** can be used to show the possible outcomes for combined events.

▶ It is particularly useful when outcomes are not equally likely.

Example

A road safety officer offered free checks on tyres and headlights for bicycles brought into school. Draw a tree diagram to show what faults a bicycle might have.

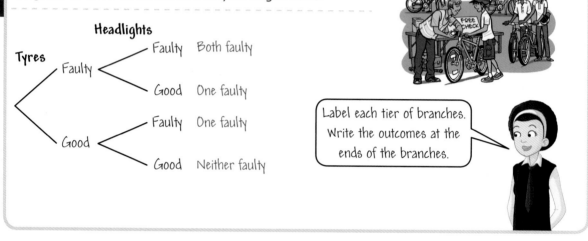

Label each tier of branches. Write the outcomes at the ends of the branches.

There is no limit to the number of possible outcomes at each stage, or how many stages there are in a tree diagram. Usually you will get two or three.

Example

Employees in a company are classified as manual, administrative or managerial and can be full-time or part-time.
Draw a tree diagram to show how employees can be classified.

There are two possible versions.

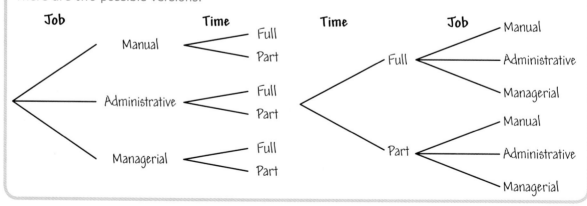

Statistics and probability Probability

Exercise 16c

1   A firm wants to classify their employees by whether they
    are married or not and whether they are male or female.
    Show a tree diagram they could use for this.

2   When James leaves the house in the morning he has a choice
    of taking an umbrella with him or not. When he comes home
    it may or may not rain. Show a tree diagram to represent this
    information.
    Which paths would result in James getting wet on his way home?

3   A football team scores 3 points for a win, 1 for a draw and 0 for
    losing a league match. Show a possibility tree for the first two
    matches of the league.
    On which paths do the team score more than 2 points altogether
    in the two matches?

Problem solving

4   On a stretch of road, a car may or may not be breaking the speed
    limit and there may or may not be a police mobile speed camera.
    a   Show a tree diagram to represent this information.
    b   Which paths would result in the car being given a ticket for
        breaking the speed limit?
    c   The police mobile camera is present one quarter of the time
        and 8% of motorists break the speed limit. If 200 cars have
        travelled along this stretch of road, show how many cars are
        on each path in your diagram.
    d   Using the number of cars on your diagram calculate the
        probability that on this stretch of road
        i   a speeding motorist is caught on camera
        ii   a speeding motorist is not caught on camera.

5   In which of the above cases do you think there is a natural order
    to the stages in the tree diagrams you drew, and in which do you
    think it could have been reversed without causing a problem?

● Tree diagrams can be used to calculate the probabilities for combinations of events.

Example

For the first example in spread **16c**, $\frac{1}{4}$ of bicycles used by teenagers do not have properly inflated tyres and $\frac{1}{6}$ of them do not have adequate lights. Show the probabilities for the various possibilities on a tree diagram.

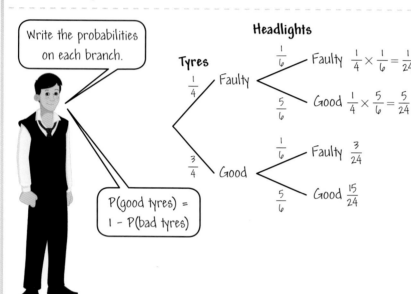

Write the probabilities on each branch.

Headlights

Tyres

$\frac{1}{4}$ Faulty

$\frac{1}{6}$ Faulty $\frac{1}{4} \times \frac{1}{6} = \frac{1}{24}$

$\frac{5}{6}$ Good $\frac{1}{4} \times \frac{5}{6} = \frac{5}{24}$

$\frac{3}{4}$ Good

$\frac{1}{6}$ Faulty $\frac{3}{24}$

$\frac{5}{6}$ Good $\frac{15}{24}$

P(good tyres) = 1 − P(bad tyres)

Assuming the two events are **independent** the individual probabilities are multiplied. Each route through the tree is a particular outcome.

● Two events are **mutually exclusive** if they cannot happen at the same time.

▶ If A and B are mutually exclusive events, then
   P(A or B) = P(A) + P(B)

For two events, 'and' is multiply; 'or' is add.

In a tree diagram the pathways are distinct so they are mutually exclusive.

Therefore, for example, the probability that a bicycle fails just one of the tests is

$$P(\text{fail one test}) = \frac{5}{24} + \frac{3}{24}$$

Faulty tyres and good headlights or good tyres and faulty headlights.

$$= \frac{8}{24}$$

$$= \frac{1}{3}$$

Exercise 16d

1 Two independent events have mutually
exclusive outcomes A, B, C and X, Y, Z
with these probabilities.

$P(A) = \frac{1}{3}$ $\qquad$ $P(B) = \frac{5}{12}$ $\qquad$ $P(C) = \frac{1}{4}$

$P(X) = \frac{1}{6}$ $\qquad$ $P(Y) = \frac{1}{3}$ $\qquad$ $P(Z) = \frac{1}{2}$

Calculate the probabilities of these
combinations of events.

a P(A and X)   **b** P(A and Y)
c P(B and Y)   **d** P(B and Z)
e P(A or B)   **f** P(A or C)
g P(B or C)   **h** P(Y or Z)
i P(X or Y)   **j** P(X or Y or Z)
k P((A and X) or (B and Y))
l P((A and X) or (A and Y))
m P((C and Z) or (B and Y))
n P((C and X) or (C and Y) or (C and Z))

Problem solving

2 Anne-Marie supports her local rugby team and goes to watch them
every week. The probability that they win a match is 0.5, that they
draw is 0.2 and the probability that it rains during the match is 0.2.
Assuming the team's performance is independent of whether it rains
or not, what is the probability that Anne-Marie watches her team
lose in the rain?

3 Colour blindness affects 8% of males. 11% of both males
and females are left-handed and this is independent of
colour-blindness. Copy and complete the tree diagram
below and calculate the probability that a male chosen
at random is left-handed and not colour-blind.

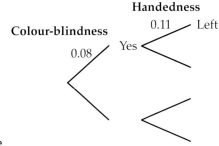

Colour-blindness   Handedness

0.08   Yes   0.11   Left

4 Eve commutes to work first taking a bus to the station
and then a train. The probability of getting a seat on the
bus is $\frac{2}{3}$ and of getting a seat on the train $\frac{3}{5}$.
a Draw a tree diagram to represent these probabilities.
b Calculate the probability that Eve has to stand on one leg of her journey.

5 For question **4** of exercise **16c**, if the probability of a win is 0.5, and the probability
of a draw is 0.3, find the probability of the team scoring more than two points from
the first two matches.

6 In question **2**, you were told that the performance
of the team is independent of whether it rains.
Do you think the outcomes of sporting events
are normally independent of the weather during
the match?
Can you find data to help you decide?

> You might take the Football league
> results for some Saturdays when the
> weather was reasonable and some
> where the weather was bad and
> compare the total number of goals
> for example.

16e Experimental probability

> Experimental probability is estimated as the fraction $\dfrac{\text{number of successes}}{\text{number of trials}}$.

Example

A road safety officer is investigating whether a busy road junction needs a cycle lane. She counted the number of cyclists arriving at the junction in each minute between 8 am and 9 am. Estimate the probabilities for the numbers of cyclists that might arrive in a randomly chosen minute.

3, 0, 2, 2, 1, 0, 4,
2, 2, 2, 1, 3, 2, 1,
0, 1, 2, 1, 4, 0, 3,
5, 2, 3, 4, 2, 1, 0,
5, 3, 2, 4, 5, 2, 4,
3, 2, 4, 5, 3, 2, 1,
0, 4, 2, 1, 3, 0, 1,
2, 4, 2, 0, 1, 3, 0,
1, 3, 4, 2

There are 60 trials.

Number of cyclists	0	1	2	3	4	5
Frequency	9	11	17	10	9	4
Estimated probability	$\frac{9}{60} = \frac{3}{20}$	$\frac{11}{60}$	$\frac{17}{60}$	$\frac{10}{60} = \frac{1}{6}$	$\frac{9}{60} = \frac{3}{20}$	$\frac{4}{60} = \frac{1}{15}$

Example

The print-out shows Sean's reaction times (in seconds) for a series of tests.

a Estimate the probability that on a randomly chosen test Sean's reaction time is less than 0.20 s.

b A prize is awarded if you can react in under 0.20 s in at least one of two attempts. Estimate the probability Sean would win a prize.

0.21	0.19	0.22	0.23	0.24	0.21
0.22	0.19	0.18	0.20	0.22	0.24
0.23	0.22	0.20	0.23	0.27	0.24
0.23	0.19	0.24	0.21	0.23	0.18
0.22	0.25	0.23	0.21	0.22	0.23
0.20	0.24	0.19	0.22	0.23	0.20
0.23	0.22	0.23	0.19	0.20	0.24
0.26	0.22	0.20	0.23	0.19	0.21
0.23	0.24	0.26	0.21	0.22	0.23
0.19	0.24	0.21	0.22	0.20	0.25

a $P(t < 0.20\text{s}) = \dfrac{\text{number of successes}}{\text{number of trials}} = \dfrac{9}{60} = 0.15$

b First attempt   Second attempt

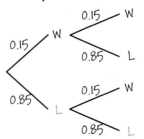

Sean will get a prize if he does not lose twice.

$P(\text{lose twice}) = 0.85 \times 0.85$

$\qquad\qquad\quad = 0.7225$

$P(\text{win at least once}) = 1 - 0.7225$

$\qquad\qquad\qquad\qquad = 0.2775$

7

Exercise 16e

1 Shamsa repeatedly tosses a fair coin until it shows heads, each time counting how many tosses it takes. The table shows her results after 40 tosses.

   a What is the relative frequency of 1s in her experiment?

   b What is probability that she will get a head on the first toss of a sequence?

   c If Shamsa was to toss the coin again, which of the answers to parts **a** and **b** would you use for the likelihood of her seeing a head on the first toss?

1	1	2	2	3
1	5	2	2	2
1	1	1	1	6
1	1	2	3	1
3	3	1	2	1
3	1	6	1	1
2	1	3	1	5
2	2	2	1	2

2 The supervisor in a factory tests samples of 25 components and records the number of faulty components in each of 30 samples. The table shows her results.

   a Find the total number of faulty components in the 750 components in these samples.

   b Estimate the probability that a component chosen at random is faulty.

0	2	1	1	0	0
2	1	2	3	1	2
1	0	1	2	0	0
3	0	0	0	1	2
1	1	0	0	0	1

Problem solving

3 The number of customers per minute arriving at the checkout in a store is recorded over a period of twenty minutes. The table shows the results listed in order.

   a Estimate the probability that three customers arrive in a minute.

   b Show that the mean number of customers arriving per minute is 2.

   c If the checkout can serve two customers per minute, make a table showing the length of the queue of customers waiting at the start of each minute.

2	1	0	3	5	4	2	3	4	2
0	3	1	2	2	0	2	1	2	1

In the first minute the two customers who arrived can be served. In the fourth minute three customers arrive so there will be one left in the queue at the start of the fifth minute.

4 In question **3**, you investigated the behaviour of the queue on the basis that two customers would be served if there were two in the queue.
What other information would let you give a better description of the behaviour of a queue?

MyMaths.co.uk   Q 1264   SEARCH

How a queue behaves over time depends on the rate of arrival at the queue and the rate of departures.

You can model the length of a queue using a **simulation**.

> new length = old length + arrivals − departures
>
> with the condition that the length can never be negative.

Consider a simple model, where the number of arrivals and departures behave according to these probability distributions.

	0	1	2	3	4
Arrivals	0.2	0.25	0.3	0.15	0.1
Departures	0.05	0.4	0.25	0.2	0.1

There is a probability of 0.2 that three people could leave the queue in a particular time interval.

To simulate 20 time periods for the queue, generate two lists of 20 **random numbers** for the arrivals and departures. Use these pairs of random numbers, $R_{\#}$ and $R'_{\#}$, with the rules given below (which use cumulative probabilities from the table above).

	0	1	2	3	4
Arrivals	$0 \leqslant R_{\#} < 0.20$	$0.20 \leqslant R_{\#} < 0.45$	$0.45 \leqslant R_{\#} < 0.75$	$0.75 \leqslant R_{\#} < 0.9$	$0.9 \leqslant R_{\#} < 1$
Departures	$0 \leqslant R'_{\#} < 0.05$	$0.05 \leqslant R'_{\#} < 0.45$	$0.45 \leqslant R'_{\#} < 0.70$	$0.70 \leqslant R'_{\#} < 0.9$	$0.9 \leqslant R'_{\#} < 1$

Running the simulation three times gave these queue histories. You can see how very different they look, even though they are all based on the same, quite simple, set of rules for a very simple situation.

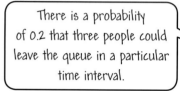

If $R_{\#} = 0.783$ and $R'_{\#} = 0.065$ there would be 3 arrivals and 1 departure in the time period.

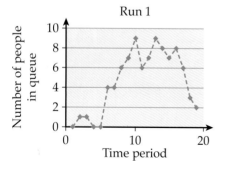

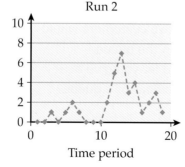

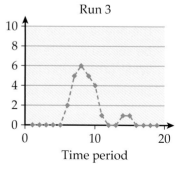

Exercise 16f

Problem solving

1 **a** Use the random number generator on your calculator , or on a spreadsheet, to simulate the arrivals and departures for 10 time periods using the probabilities in the example.

b Calculate the number of people in the queue for your results for the 10 time periods.

c Draw a graph to show your queue history, using axes like this.

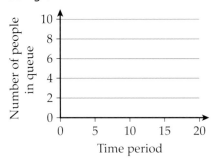

2 Repeat the simulation in question **1** and compare the two queue histories. How similar are they?

3 The store owner feels it is important that the queues do not grow to more than five, or customers will be put off using the store. A new type of till is available which gives different probabilities for the number of departures.

	0	1	2	3	4
Departures	0.05	0.2	0.4	0.15	0.2

a Assuming the arrivals continue with the same probabilities as before, use these probabilities for departures in another simulation of 10 periods.

b Would you advise the owner to get the new type of till?

4 **a** Collect your own data in a local shop to estimate probability distributions for customer arrivals and departures at the checkout and then run your own simulation.

b Use an electronic simulation to explore the behaviour of queues for a variety of probability distributions for arrivals and departures.

Did you know?

A study of motorway traffic queues led to the introduction of variable speed limits. These actually make the traffic flow faster!

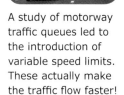
MyMaths.co.uk

16g Venn diagrams

● The **empty set**, which has the symbol ∅, contains no elements.

Ω = {1, 2, 3, 4, 5, 6, 7, 8, 9}, E = {even numbers}, F = {odd numbers} and
G = {multiples of 4}.
Find

a E ∩ G   **b** F ∪ G   **c** E ∩ F

- -

E = {2, 4, 6, 8}   F = {1, 3, 5, 7, 9}   G = {4, 8}
a E ∩ G = {4, 8}   **b** F ∪ G = {1, 3, 4, 5, 6, 8, 9}   **c** E ∩ F = ∅

● Two events are mutually exclusive if A ∩ B = ∅.

{even numbers} ∩ {odd number} = ∅

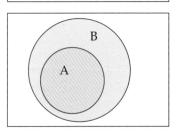

● A is a proper subset of B, A ⊂ B, if
  ▶ every element in A is also contained in set B
  ▶ set A does not contain every element in B.
● If A ⊂ B then P(A) < P(B).

{multiples of 4} ⊂ {even numbers}
All multiples of 4 are even numbers, but not all even numbers are
multiples of 4.

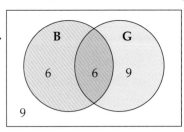

● Two events are independent if P(A ∩ B) = P(A) × P(B).

Jess surveys 30 students at her school.
She sorts them into B = {brown hair} and G = {wears glasses}.

She showed her results on a Venn diagram.
a Find P(B ∩ G).
b Prove that B and G are independent events.

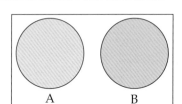

- -

a $P(B \cap G) = \frac{6}{30} = \frac{1}{5}$

b $P(B) = \frac{6+6}{30} = \frac{12}{30} = \frac{2}{5}$ $P(G) = \frac{6+9}{30} = \frac{15}{30} = \frac{1}{2}$

$P(B) \times P(G) = \frac{2}{5} \times \frac{1}{2} = \frac{2}{10} = \frac{1}{5}$

$P(B \cap G) = P(B) \times P(G)$ so the events are independent.

> Having brown hair
> doesn't affect the
> chance of someone
> wearing glasses.

Statistics and probability Probability

Exercise 16g

1 Decide if the following pairs of sets are mutually exclusive. If the sets are not mutually exclusive, state an element that is in the intersection.

   a A = {even numbers},
   B = {prime numbers}

   b C = {multiples of 3},
   D = {factors of 16}

   c E = {1, 2, 3, 5, 8},
   F = {square numbers}

1 **d** G = {multiples of 5},
   H = {multiples of 7}.

2 Decide if the following sets are a proper subset of Z = {1, 2, 3, 6, 7, 11, 12, 15, 16}. Give a reason for each answer.

   a A = {1, 2, 4}

   b B = {11, 12, 15}

   c C = {factors of 6}

   d D = {factors of 12}

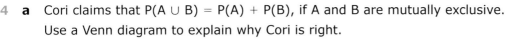

Problem solving

3 Rini sorts a group of objects into the sets A and B. She draws a Venn diagram to show her results.

   a Explain why Rini must have started with 40 objects.

   b Use $x = 2$ to find

      i P(A)   **ii** P(B′)

      iii P(A ∪ B)   **iv** P(A ∩ B)

   c If A and B are mutually exclusive, find the value of x. Hence find P(A′ ∪ B).

   d If $x = -10$, what can you say about the sets A and B?

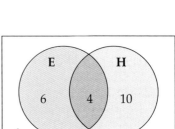

4 **a** Cori claims that P(A ∪ B) = P(A) + P(B), if A and B are mutually exclusive. Use a Venn diagram to explain why Cori is right.

   b Can you write down a formula that would work for any two events. [Hint: the formula is P(A ∪ B) = P(A) + P(B) – something]

5 Tyler sorts the 26 students in his maths class into the sets E = {blue eyes} and H = {brown hair}. His results are shown on the Venn diagram.

   a Find

      i P(E)   **ii** P(H)   **iii** P(E ∩ H)   **iv** P(E ∪ H)

   b Are E and H independent events?

   c Tyler picks a student with brown hair. What is the probability that the student has blue eyes?

   Tyler then sorts the 26 students in his maths into the sets E = {blue eyes} and G = {wears glasses}. His results are shown on the Venn diagram.

   d Are E and G independent events?

   e Tyler picks a student with glasses. What is the probability that the student has blue eyes?

   f Compare your answers to part **c** and **e** with P(E).

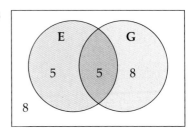

16 MySummary

Check out

You should now be able to ...

Test it ➡

Questions

✓ Evaluate uncertainty and risk in real situations.	7	1
✓ Identify and calculate probabilities for independent events.	8	2, 3
✓ Use tree diagrams to calculate probabilities.	8	4, 5
✓ Calculate probabilities from experimental data.	7	6
✓ Calculate probabilities using a Venn diagram.	8	7

Language	Meaning	Example
Independent events	Two events which have no effect on each other.	Passing your maths exam and it being a wet day could be independent events
Mutually exclusive	Two events that cannot happen at the same time.	'Toss a coin and get a head' and 'Toss a coin and get a tail'
Simulation	Using a random number generator of other software to represent an experiment.	Using your calculator random number generator to predict numbers of people in a queue.
Proper subset	B is a proper subset of A if every element of B is contained in A.	If A = {1, 2, 3) and B = {1, 2} Then B ⊂ A

1 Jon pays £5 per month to insure his mobile phone against being stolen. It would cost £240 to replace if he did not have insurance.

 a How often would Jon have to have a phone stolen in order to save money?

 b What are the advantages and disadvantages of taking the insurance?

2 Are these two events independent?
 A: the weather being rainy
 B: a person buying ice-cream

3 The probability that a person chosen at random in a shop is female is 0.8. Half the people in the shop have brown hair. Assuming these attributes are independent, what is the probability that a person chosen at random in a shop is a man with brown hair?

4 The probability of a particular type of bulb flowering is 85%. Two bulbs are planted.

 a Draw a tree diagram to show all the possible outcomes.

 b What is the probability that both bulbs flower?

 c What is the probability that exactly one bulb flowers?

5 The probability of a woman giving birth to a boy is 0.51. A women has two children. Assuming their genders are independent

 a draw a tree diagram to show the possible outcomes

 b calculate the probability of her having a boy and a girl

 c calculate the probability of her having at least one girl.

6 Aidan is practising golf on a putting green. He records how many shots he took per hole.

2	3	1	2	2
4	2	3	4	1
2	1	2	3	3

 a What is the relative frequency of 1s?

 b Estimate the probability that he will take two shots on the next hole

 c If Aidan played another 10 holes, how many times would you expect him to require more than 2 shots per hole?

7 An integer is chosen at random from the numbers 1 to 20.
 A = {multiples of 5}
 B = {triangular numbers}

 a Draw a Venn diagram to illustrate the situation.

 b Show that events A and B are not independent.

What next?

Score	0 – 3		Your knowledge of this topic is still developing. To improve look at Formative test: 3C-16; MyMaths: 1208, 1264
	4 – 6		You are gaining a secure knowledge of this topic. To improve look at InvisiPen: 463, 464, 465, 476 and 477
	7		You have mastered this topic. Well done, you are ready to progress!

16a

1. You leave a computer memory stick on a train. In the two scenarios described
   i how serious a problem is this?
   ii does it make a difference if the data is encrypted?
   a You work for a premier league football club and the stick contains records of the club's attendances over the past 5 years.
   b You work for the Inland Revenue and the stick contains the tax returns of 5837 individuals.

16b

2. State, with reasons, whether these pairs of events are independent or not.
   a i A person gets a grade C or higher in GCSE Maths.
     ii The person becomes a doctor.
   b Roll a dice and obtain
     i a factor of 6   ii an even number.

3. The probability that a car passing a checkpoint on a motorway is a sports car is 0.2. One quarter of the cars on the road are black.
   a Assuming that these are independent, what is the probability that a car chosen at random passing the checkpoint is a black sports car?
   b Can you think of any reasons why the proportions of sports cars which are black might be different to the proportion of family cars which are black?

16c

4. In tennis, a player can win or lose each set.
   a Draw a probability tree for the outcomes of the first two sets of a match.
   b On which paths are the two players level after two sets?

16d

5. The probability of Andy winning each set in a tennis match is 0.6. The match is 'best of three sets' so if a player wins the first two sets, the match is over.
   a Add probabilities to your diagram from question **4**.
   b Calculate the probability that Andy wins the match in two sets.
   c Calculate the probability that the match goes into a third set.
   d Calculate the probability that Andy wins the tennis match.
   e Why is your answer to part **d** not 0.6?

6   Arinda throws 25 darts at a dartboard, trying to score a 20.
    She succeeds on 10 of the throws.
    a   Estimate the probability of Arinda scoring a 20.
    b   Arinda throws two darts, trying to score a 20 each time.
        Draw a tree diagram to represent this.
    c   What is the probability that Arinda scores at least one 20?

7   When a dog's toy is thrown it can land blue side up or blue side down.
    Terri throws the dog's toy 320 times. It landed blue side up 64 times.
    Use these figures to estimate the probability that the next time it is thrown the
    toy lands
    a   blue side up
    b   blue side down.
    Terri throws the dog's toy 10 more times.
    It lands blue side up 4 times.
    c   Estimate the probability of the toy landing blue side up for the new 10 trials only.
    d   Add the trial results together and estimate the probability of 'blue side up' from
        the total number of trials.
    e   What is the change in the total estimated probability when the new results
        are added?

8   Use these probabilities in a queuing simulation, using the random number
    generator on your calculator, or in a spreadsheet.

	0	1	2	3	4
Arrivals	0.2	0.2	0.4	0.15	0.05
Departures	0.1	0.1	0.5	0.2	0.1

    a   Calculate the number of people in the queue for your results for the 10 time
        periods.
    b   Draw a graph to show your queue history.
    c   What was the longest queue you had over the 10 time periods?

9   Ω = {whole numbers from 1 to 20},
    T = {whole numbers from 1 to 10},
    P = {prime numbers} and C = {cube numbers}.
    a   Which set is a proper subset of T?
    b   Which sets are mutually exclusive?
    c   Prove that P and T are independent.

Forensic experts have used mathematical techniques to solve crimes for a long time. Probability, formulae and graphs are three of the topics that they need to be familiar with.

The Weekly Bugle

Following a jewellery shop raid in Park Street, Tooting, on Saturday afternoon, in which shots were fired but nothing was stolen, a getaway car was found abandoned at the junction with Fisher Row. The Ford Fiesta had narrowly missed a cyclist after skidding 53 metres. The driver and passenger of the car were seen running from the scene. A male in his 20s suspected of being the passenger was apprehended by police officers later that day. Police investigating the incident are keen to trace the driver of the car.

POLICE LINE DO NOT CROSS   PO

Task 1

Detectives searching for clues at the jewellery shop notice that a safe has been tampered with, though not successfully unlocked. The safe has a combination lock consisting of five windows that can be any one of five colours: **Green, Red, Blue, Purple or Yellow.**

Only one combination will open the safe. How many possible combinations are there?

Task 3

A DNA analysis of the abandoned car shows that two samples of DNA match the detained suspect's DNA. It is estimated that there is a one in a billion chance that a single sample of DNA will provide an exact match to another sample of DNA.

a Write the number 1 billion in standard form.
b Calculate the probability of two independent samples of DNA matching, and give your answer in standard form.
c Comment on whether or not the analysis provides evidence to support the theory that the suspect was in the car at the time of the crime.

Task 2

Detectives at the jewellery shop notice a bullet hole in a wall, at a height of 4 m from the floor. They calculate that the equation of the path of the bullet is

$$Y = x - \frac{1}{20}x^2,$$

where y is the height of the bullet above the floor in metres and x is the horizontal distance of the bullet from the gun in metres. It is believed that the person firing the shot was somewhere between 10 and 16 metres away from the wall when they fired the shot. Using trial-and-improvement, try to provide a more accurate estimate. Give your answer in metres to 1 decimal place.

CONFIDENTIAL

Task 4

The length of the tyre marks left by a skidding car depends on its speed when it started skidding.

These are typical values for a tarmac road surface and dry weather conditions.

Initial speed	length of tyre marks
10 mph	1.5 metres
15 mph	3.3 metres
20 mph	5.9 metres
25 mph	9.3 metres
30 mph	13.3 metres
35 mph	18.1 metres
40 mph	23.7 metres
45 mph	30 metres

a What happens to the length of the tyre marks as the speed doubles?

b What happens as the speed trebles?

c Is the relationship between the speed and the length of the skid a linear one?

d Use the data to draw a graph of the length of the tyre marks against speed.

e Join the points with a smooth line.

f What type of relationship does the graph show?

g Extend your graph to get an approximate speed for the car in the news article.

Task 5

The relationship between speed and the length of the skid is given by the equation

$$speed = \sqrt{90 \times length \times friction}$$

where friction is the drag factor of the road

a How far would the car have skidded if it had been on a concrete road surface with a drag factor of 0.9?

b How far would it have skidded if it had been on snow with a drag factor of 0.3?

- The tarmac road has a drag factor of 0.75

c Was the resident right in thinking that the car was doing at least 80 mph?

- You could use the equation to set up a spreadsheet.

d How quickly would it have been travelling if it skidded for 53 metres on a concrete road?

These questions will test you on your knowledge of the topics in chapters 13–16.
They give you practice in the types of questions that you may see in your GCSE exams.
There are 60 marks in total.

1   Find the nth term and the 12th term in each of these sequences
  a   7, 16, 25, 34, 43, ...   (2 marks)
  b   -2, 1, 6, 13, 22, ...   (3 marks)
  c $\frac{1}{2}$, 2, $4\frac{1}{2}$, 8, $12\frac{1}{2}$, ...   (3 marks)

2   **a**   For this prism state the number of faces, edges and vertices.   (2 marks)
    b   Draw the net of this prism.   (3 marks)
    c   How many planes of symmetry does this prism have?   (1 mark)

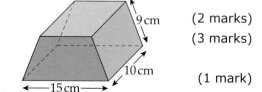

3   A simple atomic arrangement locates atoms on the corners of a cube.
    If the cube has a length of 3 nm find
    a   the distance between base atoms shown by the marked letter x (2 marks)
    b   the distance between diametrically opposite atoms shown by the marked letter y.   (3 marks)

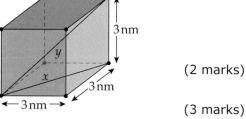

4   In the diagram find the length of the line marked m giving the answer to one decimal place.   (3 marks)

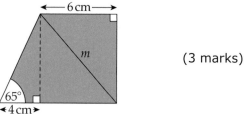

5   In the triangle show which statements are true (T) or false (F).   (6 marks)
    a $h = \dfrac{3}{\tan x}$ **b** $h = \sqrt{3^2 + 5^2}$
    c $h = \dfrac{3}{\sin y}$ **d** $h = \dfrac{\cos y}{5}$
    e $h = \dfrac{5}{\sin x}$ **f** $h = \sqrt{5^2 - 3^2}$

6   A small dinghy sails from the start position S for 0.8 km due north to a buoy at A.
    It then changes direction and sails on a bearing of 090° for 1.2 km reaching a second buoy at B. It then races back to S.
    Find
    a   the distance SB to 1 dp   (2 marks)
    b   the angle SBA   (2 marks)
    c   the bearing of S from B.   (2 marks)

7 A particular fruit drinks is made from $\frac{1}{2}$ cup of orange juice, $\frac{1}{4}$ cup
   of cranberry juice and $1\frac{2}{3}$ cup of apple juice.
   Find the ratio of apple juice to orange juice to cranberry juice,
   giving your answer in its lowest terms.   (2 marks)

8 The surface area of the Earth is 510 million km² of which $\frac{71}{100}$ is water.
   What area of the Earth is land?   (2 marks)

9 Leah earns £377 per week for doing a five day week, 8 hours per day.
   a How much does Leah earn per hour?   (2 marks)
   b How much does she earn in a year if she works 47 weeks per year?   (2 marks)

10 £12 000 is invested in a savings account as part of a pension fund.
   The current rate of interest is 5.2% per year.
   a How much is in the savings account after
      i one year   ii   three years?   (3 marks)
   b How long will it take before the savings exceed £20 000?   (3 marks)

11 A 3-sided spinner has sections coloured red, blue and green
   with sector areas $\frac{1}{4}$, $\frac{1}{8}$ and $\frac{5}{8}$, respectively. The game involves
   two spins of the spinner.
   a Draw a tree diagram to represent the probabilities.   (3 marks)
   b What is the probability of obtaining
      i   consecutive red sections   (1 mark)
      ii   a red-green combination   (2 marks)
      iii the same colour on consecutive spins?   (2 marks)

12 The number of breakdowns per week recorded by a taxi firm
   during a six month period is shown.

```
2 1 0 0 1 2 2 3 0
3 2 1 1 0 0 0 2 1
4 1 0 1 2 3 1 0 0
```

   The taxi company runs 25 taxis.
   Find
   a The total number of breakdowns in this six-month period   (1 mark)
   b An estimate of the probability that one or more taxis will
      breakdown during any week.   (3 marks)

You are about to meet six students who have volunteered to go to Kangera in East Africa to work with a charity helping to rebuild a school. There is a lot of work to be done and your mathematical skills will be put to good use helping them.

Solving real life problems often requires you to think for yourself and to use several pieces of mathematics at once. If you are going to be successful you will need to practice your basic skills.

- **Fluency** – Can you confidently work with numbers, graphs and formulae?

- **Reasoning** – Can you use algebra to describe a situation?

- **Problem solving** – Can you cope if a problem doesn't look familiar?

Fluency

Being good at mathematics doesn't just mean being able to do sums and solve equations.

It can also mean how you think about things. Can you look at a 2D drawing and see the 3D shape?

Scale drawings, based on careful measurement and 3D reasoning, are used in architecture and engineering to plan projects. Many of the same ideas are used to produce the graphics in computer games.

Reasoning

It is often said that we are living in the information age. Newspapers, television, the internet… they are all bombarding us with statistics and they are not always right.

To making sense of all this information you need to understand the mathematics and be able to reflect on what the statements might mean.

Perhaps you will be asked to produce a report yourself explaining some statistics.

Problem solving

Scientists use mathematics to describe the world around us. By understanding how forces change on a scale model, engineers can investigate the drag on a real car.

They can then use real and mathematical models to ensure that cars are more fuel-efficient.

The AfriLinks project links schools in Europe with schools in Africa.

Six British students travel to Kangera in East Africa.

The local school has been destroyed in a mudslide. The students will help build a new school.

Greg

Ella

Imran

Maxine

Josh

Wah Wah

Before they depart, the students find out about the Kangera region.

1 The population is 9953.

The table shows the population by age

Age group	0 – 20	21 – 40	41 – 60	60+
Number	3490	2988	2450	1025

a Round these numbers to the nearest 100.

b Draw a pie chart to display the rounded data values.
Make sure you provide a key.

c To the nearest whole number, what percentage of the population is over 60?

2 The graph shows average temperature and rainfall over a 30 year period in the Kangera region.

a What is the range of temperatures?

b What is the range of rainfall measurements?

c What is the median average rainfall for a year?

d What is the mean average temperature for a year?

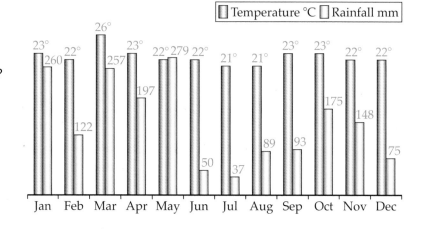

Everyday maths

3 The table shows rainfall measurements
for the first five months of this year.

Month	Jan	Feb	Mar	Apr	May
Rainfall (mm)	180	242	368	492	481

  a What do you notice when you compare these figures with
    the average rainfall?

  b Explain why you think the school building was destroyed.

Most pupils walk to school in Kangera.
The map shows the journey for some of the pupils.

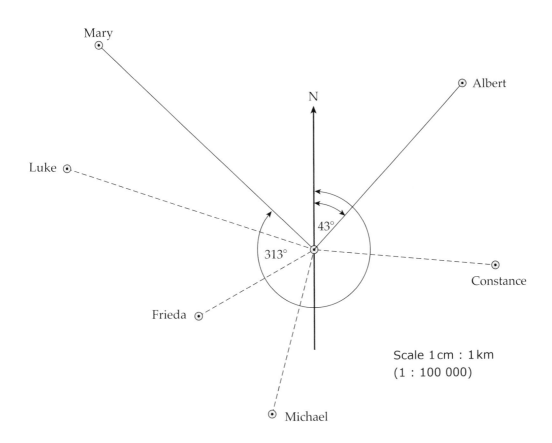

Scale 1 cm : 1 km
(1 : 100 000)

4 a Give the three-figure bearings and distances for the journeys
    from school to home for Constance, Michael, Frieda and Luke.

    From school Albert's home is 6 km away on a bearing 043°.
    From school Mary's home is 8 km away on a bearing 313°.

Work with a partner to solve these problems.

  b What is the bearing from their homes to school for

    i Albert   **ii** Mary?

  c What is the distance between Mary's and Albert's homes?

  d What is the bearing of Albert's home from Mary's home?

MyMaths.co.uk

This is a draft plan of the schoolhouse. It is not drawn to scale.
The plan shows the plan view, side elevation and front elevation.

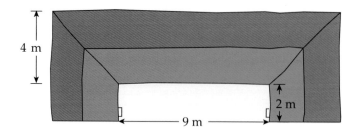

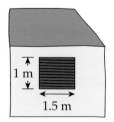

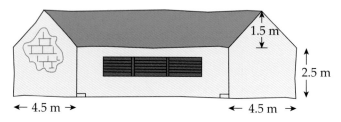

1   a   Sketch a 3D drawing of the schoolhouse.
        Start like this.

    b   Label the sketch to show these measurements

        i height   ii length   iii total width.

Ella buys the breeze-blocks for building the walls.
Each block measures 20 cm × 50 cm × 20 cm.

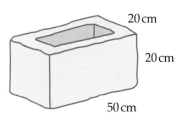

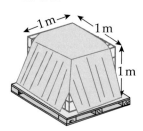

2   The blocks are sold by the pallet.
    The stack of blocks measures 1 m × 1 m × 1 m.
    How many blocks are there on each pallet?

3   A wooden pallet weighs 28 kg.
    Each block weighs 4.8 kg.
    A pick-up truck can just about carry a load of two tonnes.
    Maxine thinks that it will be able to carry 8 pallets at a time.
    Show that her calculations are wrong.

4   The gable walls are covered in a cement mixture.

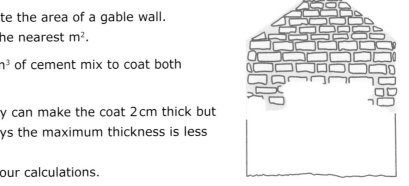

   a   Use the plans to calculate the area of a gable wall.
        Round your answer to the nearest m^2.

   b   The students have 0.5 m^3 of cement mix to coat both
        gable walls.

        Josh works out that they can make the coat 2 cm thick but
        Imran disagrees and says the maximum thickness is less
        than that.
        Who is correct? Show your calculations.

5   Maxine notices that there seem to be quite a few cracked or
    broken breeze–blocks. As Imran lays the blocks Maxine records
    the number of damaged blocks that they come across.
    She samples every hundred blocks laid.

Sample	1	2	3	4	5	6	7	8	9	10
Cracked	3	11	3	0	21	1	12	13	0	23
Broken	6	0	3	0	5	6	1	8	0	8
Good	91	89	94	100	74	93	87	79	100	69

   a   Maxine estimates that 70% of the blocks are in good condition.
        Imran estimates that nearer to 90% are in good condition.
        Use the table to see who is closest. Show all of your working.

   b   Use the table to estimate how many blocks Imran can expect
        to be damaged if they ordered 6000 blocks to build the school.

   c   If they need 6000 undamaged blocks how many should
        they order?

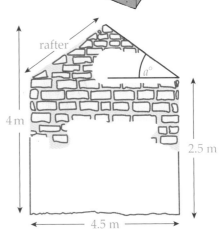

6   **a**   Calculate the length of the rafters of the roof.

   b   To cope with heavy rainstorms, the roof must
        have a 'pitch' or angle of at least 35°.
        Will the schoolhouse roof will be safe in the rain?

Greg and Wah Wah are laying the path.

1 A pallet of paving blocks is delivered.
The surface of each block is 10 cm by 20 cm.

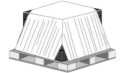

10 cm

20 cm

There are 75 blocks on a pallet.

a What area will one pallet cover?

b They have to cover 7 m².
How many pallets will they need to order?

They try out some different patterns for the paths.

2 This is Wah Wah's pattern.

She counts the tiles and says they have 8 yellow tiles
and 160 red tiles.

If they use all the yellow tiles, how many red tiles will
they have left over?

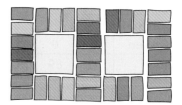

3 Greg has his own idea using red blocks and green slabs.

BLOCK

20 cm

SLAB

750 mm

a How many green slabs will need to be laid before the
pattern begins to repeat itself?

b If the path is 12 m long, how many times will the pattern
repeat?

c How many blocks and slabs will need to
be bought?

d What percentage of the path is made up of the green slabs?

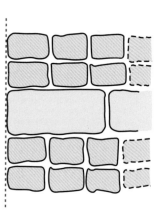

4 Greg and Wah Wah notice that this basic
shape on Jacob's shirt tessellates
to form this pattern.

Using isometric paper, make your own copy of the shape
and show that it tessellates.

5 Wah Wah finds a catalogue containing paving stones in the shape of
regular polygons.

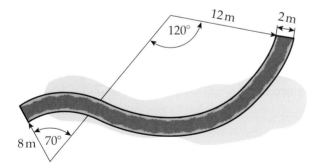

a Greg says only three of these shapes will tessellate, is he right?

b Wah Wah says she can make a tessellation using a pair of shapes.
Sketch a pattern that she could use.

c Is it possible to make a tessellation using three of the shapes?

Remember that
the angles at a
point must add
up to 360°.

6 Maxine is planning another path made from arcs of circles
with differing radii.

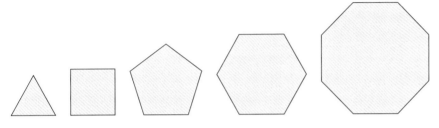

a The path is 2 m wide, so she bases her calculations on the centre line of the path.
What is the length of the path?

b The cost of the path is $10.25 per metre for the first 5 m and $9.50 for the rest.
To the nearest $5.00 what is the total cost?

The students are going to help make a basketball court.

1   Josh makes the basket by bending a metal strip into a circle.

 22.5 cm

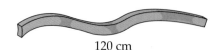 120 cm

The radius of the ring has to be 22.5 cm.
Josh makes a mistake and only cuts a 120 cm length of metal strip.

a   What is the radius of the circle made by this strip?

b   Calculate the percentage error in the cutting of the strip.

Use π = 3.14

Maxine straightens the strip and bends it into a ring of the correct radius.

   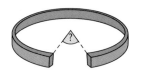

With a radius of 22.5 cm, the strip is too short and makes only an arc, not the complete circle.

c   To the nearest 5°, what angle is made by the arc of the circle?

2   Ella sinks the basket poles into wet concrete.

To support them whilst the concrete dries she anchors them to the ground with 6 m long wires. She connects the wires 3 m up the pole.

a   To the nearest 10 cm calculate the distance of the anchor from the pole.

b   What angle do the wires make with the ground?

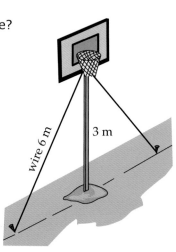

wire 6 m   3 m

3   The students have a school team. They are asked to choose their own kit from these colours.

a   Greg says that there are 13 different combinations of kit using these colours.
Draw a probability table and hence explain why he is wrong.

b   What is the probability that the kit will be all the same colour?

When they have completed the basketball court the students organise a celebratory tournament.

They play the local school team.
Here are the records of their performances.

Number of baskets per game	Interval mid-point	Local team			Visiting team		
		Frequency	Cumulative frequency	Estimated total	Frequency	Cumulative frequency	Estimated total
0 to 4	2	0	0	0	2		
5 to 9	7	1	1	7	1		
10 to 14		1	2		3		
15 to 19		1			3		
20 to 24		2			6		
25 to 29		2			5		
30 to 34		3			5		
35 to 39		8			4		
40 to 44		9			1		
45 to 49		3			0		
	Totals						

4 **a** Copy and complete the above table and the accompanying cumulative frequency graph.

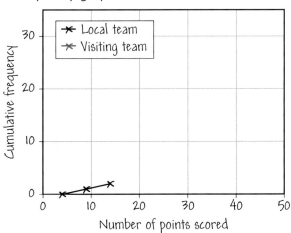

	Local team	Visiting team
Total number of games played		
Modal class		
Estimated Mean		
Median score		

b Using the above results and graph complete the table of statistics.

c Using the three averages, which team was the stronger side?

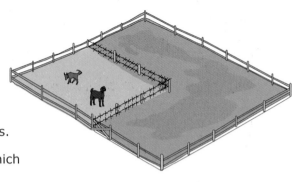

The school helps to grow food for the village.
They fence off a plot of land 20 m × 20 m.

The head teacher says he wants to divide the
field into five sections.

One quarter of the field is reserved for goats.
The rest is divided up into four congruent shapes.

1 a Work with a partner to find the shape which
will tessellate for the four crops.

b Make a sketch of the shape and show its
measurements.

c What is the area given to each crop?

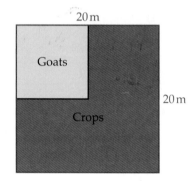

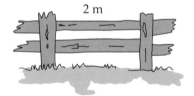

2 The field is bounded by a fence.
Greg drives posts into the ground 2 m apart and
joins them with two cross-bars.

a Greg has ordered 38 posts to fence the large field.
Is this the right number?
Show your calculations.

b Construct a formula to connect the length of fencing in
metres and the number posts needed.
Use *p* for the number of posts used and *m* for the length
of fencing.

c Show that your formula works for part **a**.

3 Greg, Maxine and Josh start to construct the fence.
They estimate that it will take them ten days.
This is too long to wait; the head teacher needs the fence
completed in four days.

a How many more workers will be needed to grant the head
teacher his wish?

b What does your calculation assume?

4 Here is another rectangular plot of land which will be used for keeping hens safely.
Its area is $36\,m^2$.
One of its sides has to be 7m shorter than the other.

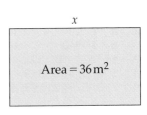

x

Area $= 36\,m^2$

a If the longer side is x metres long, write an expression using x for the shorter side.

b Hence construct an expression for the area of the field.

c Solve the equation and use your answer to find the dimensions of the plot.

5 Josh and Wah Wah are preparing a planter for a flower display.
The shape of the planter is a prism with a trapezoidal cross-section.

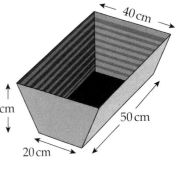

40 cm

20 cm

50 cm

20 cm

a Find the volume of the planter.

b The compost is delivered in a 72 litre bag.
When the planter is filled, what percentage of the compost remains?

6 Imran puts 3 bags of soil and 2 bags of compost onto the weighing scales and then sits on top of the bags.
The total weight is 146 kg.

Imran puts 4 bags of soil and 4 bags of compost onto the weighing scales and then sits on top of the bags.
The total weight is 195 kg.

Imran puts 1 bag of soil and 3 bags of compost onto the weighing scales and then sits on top of the bags.
The total weight is 123 kg.

Calculate the weights of the two types of bags and Imran.

Check in 1

1 a i 10   ii 13.0   iii 12.973
   b i 340   ii 342.9   iii 342.914
2 a 27 b 32 c 1 000 000 d 625
3 1, 2, 3, 4, 6, 8, 9, 12, 18, 24, 36, 72
4 2, 3, 5, 7, 11, 13, 17, 19, 23, 29

MyReview 1

1 a 3   b 3.1   c 3.08
2 a 0.04   b 0.036   c 0.0360
3 a 6   b 50   c 1000   d 50
4 a 1.5 cm, 0.5 cm
   b 1.505 m, 1.495 m
   c 25 mm, 35 mm
   d 245 mm, 255 mm
5 a 100.5 mm, 99.5 mm
   b 8.05 cm, 7.95 cm
   c 10.95 cm, 11.05 cm
   d 0.5 mm, 1.5 mm
6 £485 million ≤ cost < £495 million
7 a 29.75 cm²   b 20 cm
8 a 25.2 m/s   b 26.1 m/s
9 a $2^2 \times 3 \times 7$ b $2 \times 3^2 \times 7$
   c $3^2 \times 7 \times 11$ d $2 \times 5 \times 11 \times 13$
10 a HCF = 5, LCM = 175
   b HCF = 6, LCM = 2574
   c HCF = 105, LCM = 1260
   d HCF = 60, LCM = 25200

11 a $\dfrac{135}{420} = \dfrac{9}{28}$ b $\dfrac{24}{420} = \dfrac{2}{35}$

Check in 2

1 a 1430   b 35   c 36.1
   d 300   e 6000
2 a 37.68 cm   b 113.04 cm²
3 a 236 cm²   b 240 cm³

MyReview 2

1 a 6820 mm   b 0.14 l
   c 820 g   d 70 mm²
   e 80 000 cm²   f 12 000 cm³
2 a 4.8 l   b 10 kg
   c 192 km   d 0.625 m
3 a area b volume c length d area
4 390 cm²
5 11 cm
6 a 283.5 cm²   b 59.7 cm
7 a 36π cm²   b 24 + 6π cm
8 3.2 g/cm³
9 a 59.5 km   b 12 minutes

Check in 3

1 a 3x   b 9y   c 7a
   d 5b − 2   e 4p   f a + 5b
   g x² + 2x   h 6y + y²
2 a 42 cm   b 46 cm
3 a 2x + 10   b 7p − 21   c 10k + 5
   d 12 − 3y   e x² + 2x   f ab − 6a
   g 4p² + 4pq   h 15t − 6t²

314 Answers

MyReview 3

1 a 1000   b -8
2 a 20a⁸   b 3b⁵   c 27c¹⁸   d 2d⁶
3 a 1   b 4   c $\dfrac{1}{12}$ d $\dfrac{1}{10}$
4 a e   b $f^{\frac{3}{2}}$ c g⁵   d h
5 a a² + 3a + 2   b b² + 3b − 28
   c c² − 11c + 24   d d² − 10d + 25
   e 2e² − 5e − 12   f 9f − 36
6 a 4(g − 5)   b 8(4 − 3h)
   c 5m(3n + 4)   d 4y(3x + 1)
7 a x(x + 7)   b 5x(x + 2y)
   c 2y(2y² + 3y − 1)   d 3y²(4y − 5z)
8 a (p + 3)(p − 3)   b (2q + 1)(2q − 1)
   c (r + 9s)(r − 9s)   d (10t + u)(10t − u)
9 a yes   b no   c yes
10 2a² + 30a − πa² / a(2a + 30 − πa)

11 a x = b − 2a   b $x = \dfrac{c + b}{a}$

   c x = 3f(d + e) d $x = \dfrac{c − 2a}{b}$

12 a $y = \dfrac{a}{2b}$ b $y = \dfrac{3}{c} − d$

   c $y = \dfrac{2}{f − e}$ d $y = \sqrt{\dfrac{a − 3c}{b}}$

Check in 4

1 a $\dfrac{13}{21}$ b $\dfrac{6}{35}$ c $\dfrac{8}{15}$ d $\dfrac{5}{6}$
2 a £36   b 16.8 kg c £64.80 d £337.75
3 a 3 : 4   b 13 : 5   c 7 : 30   d 2 : 3

MyReview 4

1 a $\dfrac{171}{35}$ b $\dfrac{91}{36}$ c $\dfrac{77}{20}$ d $\dfrac{107}{60}$
2 a $\dfrac{7}{3}$ b $\dfrac{3}{20}$ c 18
   d $\dfrac{8}{5}$ or $1\dfrac{3}{5}$ e $\dfrac{5}{2}$ or $2\dfrac{1}{2}$ f $\dfrac{7}{2}$ or $3\dfrac{1}{2}$
3 a 0.7̇   b 0.45̇   c 0.416̇
   d 0.857142̇   e 0.63̇   f 0.538461̇
4 a $\dfrac{31}{99}$ b $\dfrac{1}{15}$
5 a £94.05   b 485.55 g
6 6.67%
7 30.6%
8 £720
9 1.70 m
10 £749.42
11 £7111.11

Check in 5

1 a 93°   b 27°   c 56°
2 a i 31.4 cm   ii 78.5 cm²
   b i 37.7 m   ii 113.1 m²
   c i 0.63 m   ii 0.031 m²

MyReview 5

1 a = 12°, angles in a triangle add up to 180°.
   b = 103°, corresponding angles are equal.

$c = 106°$, missing angles in triangle are both 74° as isosceles then use angles on a straight line adding up to 180°.

$d = 72°$, missing angles in triangle are both 54° as isosceles.

$e = 109°$, use alternate angles are equal to find angle adjacent to e then angles on a straight line.

$f = 98°$, angles on a straight line.

$g = 82°$, either use angles in polygon add up to 360° or alternate angles are equal.

2 **a** Exterior = 60, interior = 120°
 b Exterior = 20, interior = 160°

3 **a** 45 **b** 172° **c** 7740°
 d 45 **e** 45

4 **a** 55° **b** 55° **c** 35°

5 **a** $\dfrac{40}{9}\pi$ **b** $\dfrac{160}{9}\pi$

6 7.78 cm

7 **a** $p = 65°$, $q = 51°$ **b** Congruent

Check in 6

1 **a** 3 **b** 0 **c** -6 **d** 15

2 **a** Straight line through (0, 1)
 b Straight line through (0, 5)
 c Straight line through (0, 12)
 d Straight line through $(0, 3\frac{1}{3})$

MyReview 6

1 $\dfrac{1}{2}$

2 **a** $m = \dfrac{1}{4}$, $c = 0$ **b** $m = -7$, $c = 10$
 c $m = 8$, $c = -28$ **d** $m = -4$, $c = 5$

3 **a** $y = 2x - 3$ **b** $y = 4 - x$
 c $y = \dfrac{x}{2} + 1$

4 $y = 3x \pm c$

5 **a**

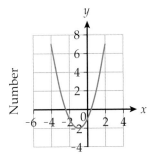

Number of items of homework

 b $x = -1$ **c** (-1, -2)

6 **a**

x	-3	-2	-1	0	1	2	3
$4x$	-12	-8	-4	0	4	8	12
$-x^3$	27	8	1	0	-1	-8	-27
y	15	0	-3	0	3	0	-15

b

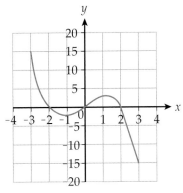

7

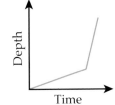

8

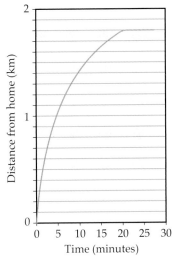

9 **a** Reciprocal graph, asymptotes $x = 0$ and $y = 0$. Passing through (-1, -4) and (1, 4). $x \approx 2.1$
 b Exponential graph, passes through (0, 1) and (1, 3). $x \approx 2$

Check in 7

1 **a** 688.8 **b** 148.58
 c 23.5 **d** 45.9

2 **a** 4.472 **b** It was rounded

MyReview 7

1 **a** 148.02 **b** 4.15
 c 15.68 **d** 17.58
 e 9.21 **f** -88.35

2 **a** 94.963 **b** 129.88
 c 8.758 **d** 44.2411

3 **a** 182.16   **b** 34.79
     c 4.524   **d** 0.031584
4 **a** 120.12   **b** 4.108
     c 22.272   **d** 7.2226
5 **a** 132   **b** 689   **c** 6950
     d 0.0268   **e** 4.99
6 **a** 0.88   **b** 0.72
7 **a** 345 000 000 km
     b 439 days, 16 hours and 36 minutes
8 **a** £7.14   **b** 33 mm
     c 19 hours 12 minutes

Check in 8
1 **a** 0   **b** 2   **c** 7 − 0 = 7   **d** 2.3
2 **a** Own diagram   **b** $25 \le x < 30$

MyReview 8
1 **a** 19   **b** 5 − 9
     c

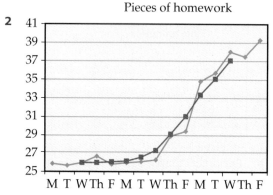

2

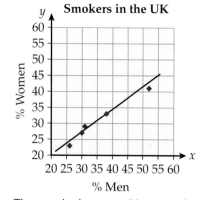

3 9.17 words
4

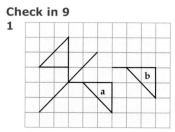

The graph shows positive correlation.

5 **a**

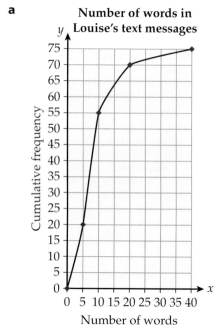

b 7.5
c IQR = 11 − 4.5 = 6.5

6

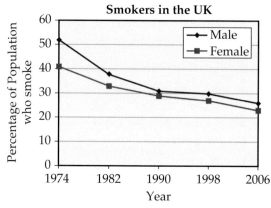

7

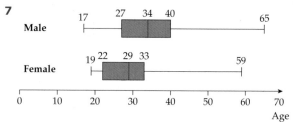

Check in 9
1

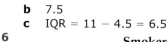

2 195 m, 0.8 m

MyReview 9
1 a, b, c

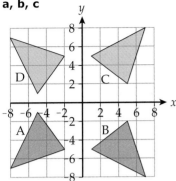

d Reflection in line $y = 0$

2 a $\begin{pmatrix} 1 \\ -7 \end{pmatrix}$ **b** (-8, -4)

3

4 Scale factor $\frac{1}{3}$, centre of enlargement (0, 9)

5

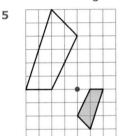

6 a

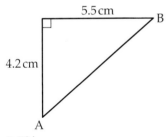

5.5 cm
B
4.2 cm
A

b 5.52 km
7 2.2 cm

Check in 10
1 a $24x + 120$ **b** $20a + 27$
   c $5x^2 - 20x$ **d** $10y - 5$
2 a $x = 3$ **b** $m = 7$
3 $x^2 + x + 5$

MyReview 10
1 a $a = 13$ **b** $b = 21$
   c $c = \frac{3}{2}$ **d** $d = -4$
   e $e = -\frac{2}{5}$ **f** $f = \frac{4}{7}$

2 a $x = 2, y = 7$ **b** $x = -1, y = 6$
   c $x = \frac{1}{2}, y = \frac{3}{2}$ **d** $x = -4, y = \frac{1}{2}$
3 a $x = 5, y = -2$ **b** $x = -4, y = 7$
   c $x = -6, y = -9$ **d** $x = 4, y = -\frac{1}{2}$
4 a $x = 6, y = 3$ **b** $x = -2, y = 6$
   c $x = \frac{1}{2}, y = -\frac{1}{2}$ **d** $x = \frac{2}{5}, y = \frac{6}{13}$
5 a $s = 14$ ($m = 39$)
6 a

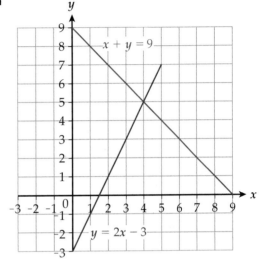

$x = 4, y = 5$

b

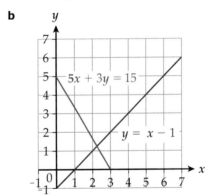

$x = 6, y = 5$

7 a $x \leq -1$ **b** $x < 6$

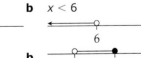

8 a 5, 6, 7   **b**

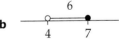

9 a 2.78   **b** 1.46

Check in 11
1 a 0.038   **b** 27.5   **c** 0.045   **d** 100
2 a 17   **b** 0.75   **c** 2000   **d** 0.040

MyMaths.co.uk

MyReview 11
1 **a** 21 000 000 **b** 269 000
  c 0.0074 **d** 0.000373
2 **a** 2.6×10^4 **b** 9.88×10^5
  c 8.8×10^{-5} **d** 7.59×10^{-3}
  e 5.94×10^9 **f** 9.8×10^{-9}
  g 1.4×10^4 **h** 6.72×10^{-2}
3 **a** 8.495×10^6 **b** 5.172×10^{-2}
4 **a** 9^8 **b** 8^{-9} **c** 6^7
  d 3^{-11} **e** 4^{14} **f** 7^{15}
  g 3^{12} **h** 4^{-12}
5 **a** 7 **b** $5\sqrt{2}$ **c** $6\sqrt{6}$
  d $4\sqrt{2}$ **e** $24\sqrt{10}$ **f** $48\sqrt{3}$
6 **a** $4\sqrt{2}$ **b** $4\sqrt{5}$
  c $5\sqrt{7}$ **d** $8\sqrt{5}$
7 **a** $7^{\frac{1}{2}}$ **b** $5^{\frac{1}{3}}$
8 **a** 7 **b** 4
9 **a** 6^7 **b** $7^{\frac{3}{2}}$ **c** 1
  d $8^{\frac{7}{2}}$ **e** $4^{\frac{2}{3}}$ **f** $3^{\frac{3}{2}}$

Check in 12
1 **a** 17 **b** 12 **c** 5.67 **d** 5.4
2 **a** Check BC = 6 cm
  b Check that perpendicular bisector passes
    through C.
  c **i**, **ii**, **iii** all 3.5 cm.

MyReview 12
1 **a** 26 cm **b** 8 m **c** 15 mm
2 No
3 **a** $5\sqrt{5}$ **b** $4\sqrt{17}$
4 6.71 m
5 $5\sqrt{2}$
6 Check ASA: 48°, 5 cm, 53° for top triangle and
  SSS: 5 cm, 7 cm, 4 cm for bottom triangle
7 Check ASS 25°, 5 cm, 3 cm for both triangles.
  Solution 1: AC = 2.4 cm, angle B = 20°,
  angle C = 135°.
  Solution 2: AC = 6.7 cm, angle B = 110°,
  angle C = 45°.

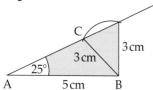

8 FORWARD 4
  REPEAT 2
  [TURN RIGHT 120° FORWARD 4]
9

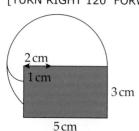

Check in 13
1 **a** Start at 3 and +3; 12
  b Start at 2 and +5; 12
  c Start at 4 and +3; 10, 13
  d Start at 21 and -4; 17, 13
  e Start at 5 and +8; 13, 29
  f Start at 50 and -6; 44, 32
2 **a** 0, 3, 6, 9, 12 **b** 1, 2, 4, 8, 16
  c 0, -2, -4, -6, -8 **d** 2, 1, $\frac{1}{2}$, $\frac{1}{4}$, $\frac{1}{8}$

MyReview 13
1 **a** $5n - 4$ **b** $7n - 3$
  c $0.9n + 4.7$ **d** $20 - 2n$
2 **a** 3, 9, 15, 21, 27
  b -4.5, -4, -3.5, -3, -2.5
  c 28, 52, 76, 100, 124
  d 19, 16, 13, 10, 7
3 **a** 465 **b** 820
4 **a** $n(n + 2)/n^2 + 2n$ **b** 120
5 **a** 2, 5, 10, 17, 26 **b** 4, 16, 36, 64, 100
  c -3, 6, 27, 60, 105 **d** 5, 11, 19, 29, 41
6 **a** $n^2 + 5$ **b** $2n^2 + 3n$
  c $4n^2 - 2n + 1$ **d** $n^2 - 5n + 2$
7 **a** Divergent
  b Converges to a limit of ⁻2
  c Divergent
  d Oscillatory

Check in 14
1 **a** Cylinder **b** Pyramid
  c Prism **d** Cone
2 **a** 2.5 cm **b** 24 cm

MyReview 14
1 **a** 6 faces, 12 edges, 8 vertices
  b

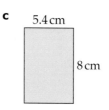

  c 5.4 cm, 8 cm

2 **a** 15 cm **b** 1512 cm³ **c** 1116 cm²
3 **a** $4\sqrt{5}$ cm **b** $4\sqrt{6}$ cm
4 sin θ = 0.385, cos θ = 0.923, tan θ = 0.417

5	a	36.9°	b	78.5°	c	54.5°
6	a	33.1°	b	66.4°	c	34.6°
7	a	7.1 cm	b	9.7 cm	c	22.8 cm
8	a	12.5 km				

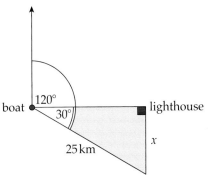

b 300°

Check in 15

1	a	£235.84	b	43.3125 kg		
2	a	8.93%	b	1.47		
3	a	30	b	0.08	c	3000
	d	4×10^2	e	5×10^{-2}		

MyReview 15

1	a	£6$\frac{3}{4}$	b	£6$\frac{1}{3}$ miles
2	a	£21$\frac{2}{3}$	b	960 kg
3	a	27 km	b	£54$\frac{1}{3}$ m
4	a	4 : 2 : 5	b	13 : 8
5	a	1 : 3.5	b	1 : 3.75
6	The first one		
7	a	210	b	78	c	8
8	£1.47		
9	£5.85		
10	a	1.05 km / 1050 m	b	30 cm
11	a	7.143 litres	b	2.679 litres
12	Bottle C.		

Check in 16

1	a	$\frac{1}{3}$	b	$\frac{2}{3}$		
2	a	$\frac{3}{4}$	b	$\frac{3}{8}$	c	1
	d	$\frac{1}{9}$	e	$\frac{14}{25}$	f	$\frac{61}{36}$
	g	$\frac{1}{3}$	h	$\frac{7}{20}$		

MyReview 16

1 a Every 4 years
  b Advantages: peace of mind, may not be
    able to afford £240 in one go.
    Disadvantages: could end up paying a lot
    of money and never need the insurance.

2 No
3 0.1 or 10%
4 a

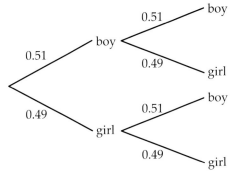

b 72.25% or 0.7225   **c** 25.5% or 0.255

5 a

boy / girl probability tree with values 0.51 and 0.49

b 0.4998 or 49.98%   **c** 0.7399 or 73.99%
6 a   0.2   **b**   0.4   **c**   4
7 a

Venn diagram: A and B. A only = 2, intersection = 2, B only = 3, outside = 13

b $P(A) = \frac{1}{5}$, $P(B) = \frac{1}{4}$, $P(A) \times P(B) = \frac{1}{20}$
$P(A \cdot B) = \frac{1}{10}$, $P(A \cdot B) \cdot P(A) \times P(B)$
so A and B are not independent.

Index

reflection symmetry 76, 244
regular polygon 76, 78
remainder 124
repeated percentage change 64, 274
response variable 132
Retail Price Index (RPI) 138
reverse bearing 252
reverse percentages 62
RHS triangle 82, 216
right-angled triangle 212, 214, 246,
   248, 250
risk 280
rotation 160
rounding 4, 6, 12
rounding errors 122, 124, 126

S
salary 273
sample 134, 152
SAS triangle 82, 216
scale 166
   proportion and 268
scale drawing 166, 170
scale factor 162, 168, 170, 268, 274
   negative 164
seasonal variation 106
second difference 234
secondary data 132
sector 80, 84
self-similar object 169
sequence 230
   behaviour of 236
   convergent 236, 238
   divergent 236, 238
   oscillatory 236, 238
   quadratic 234, 238
significant figures (sf) 4, 12
similar shapes 162, 168, 170
simulation 290, 294
simultaneous equations 180, 182, 192
   construction of 184
   solving with graphs 186
sine (sin) 248, 250, 254
sketch graph 104
solid
   net of 244
   reflection symmetry 244
solution 190
solution set 188, 192

speed 24, 26, 102
   average 24
spread 148
square numbers 232
square root 46, 204, 212
SSA triangle 216
SSS triangle 82, 216
standard index form 10, 32, 196, 206
   cube roots and 204
   division and 202
   for large numbers 198
   multiplication and 202
   for small numbers 200
   square roots and 204
   use of calculator 122
stationary 102
statistical survey 132
straight line
   gradient of 90
   graph 92, 104, 110
subject of formula, changing 44, 46, 48
substitution 42
subtraction
   of decimals 120
   of fractions 56
   of indices 32, 34, 202
   as inverse of addition 44
   order of operations 122, 126, 202,
      206
surd 204, 206
surface area 22, 246
survey 132, 134
symmetry 76

T
tangent (tan) 248, 250, 254, 258
tangent to circle 78, 84
term 178
terminating decimal 58, 66
tessellation 76
theta (θ) 80, 248
three-figure bearing 252
time series 106
time series graph 110
transformation 160
translation 160
trapezium 23
   area of 22
tree diagram 284

trend 106, 146
trend line 138
trial-and-improvement 190
triangle
   AAA 216
   area of 22
   ASA 82, 216
   congruent 82
   construction of 216
   equilateral 214
   isosceles 184
   RHS 82, 216
   right-angled 212, 214, 246, 248, 250
   SAS 82, 216
   similar 248
   SSS 82, 216
   unique 216
triangular numbers 232, 238
triangular prism 22
trigonometric ratio 250
trigonometry 248, 250

U
uncertainty 280
unique triangle 216
unitary method 272
units 24
unknown value 40, 184
upper bounds 6, 8, 12
upper quartile 144

V
value for money 272
variable 40, 42, 48, 180, 182
   control 132
   explanatory 132
   response 132
Venn diagram 10, 12, 292
volume 18, 20, 26
   of cylinder 47
   of prism 246

Y
y-intercept 92